MODERN CONCEPTS IN CHEMISTRY

EDITORS

Bryce Crawford, Jr., University of Minnesota
W. D. McElroy, Johns Hopkins University
Charles C. Price, University of Pennsylvania

WARREN L. REYNOLDS, Ph.D., University of Minnesota, is Associate Professor of Inorganic Chemistry at the University of Minnesota, where he has been associated with the faculty of the Department of Chemistry since 1955. Dr. Reynolds' special fields of interest include oxidation-reduction and substitution reactions in inorganic chemistry, stability and lability of complex ions, and bonding in inorganic species. His published papers have appeared in *Inorganic Chemistry, Journal of Chemical Physics, Journal of Physical Chemistry, Transactions of the Faraday Society,* and *Talanta.*

RUFUS W. LUMRY, Ph.D., Harvard University, is Professor of Physical Chemistry at the University of Minnesota. He is also Chairman of the Graduate Biophysics Program and Director of the United States Atomic Energy Commission Training Program for Photochemistry and Energy Transfer for Biological Scientists. Dr. Lumry is author or co-author of some 70 papers in his fields of interest, which include electron-transfer reactions and energy-transfer mechanisms, enzymes and proteins, and chemiluminescence.

MECHANISMS

OF

ELECTRON TRANSFER

WARREN L. REYNOLDS
AND
RUFUS W. LUMRY

BOTH UNIVERSITY OF MINNESOTA

THE RONALD PRESS COMPANY • NEW YORK

Library of Congress Catalog Card Number: 66–20087
PRINTED IN THE UNITED STATES OF AMERICA

Preface

This volume presents an introduction to what are now thought to be the principal mechanisms of electron-transfer reactions. We have hoped to make it suitable for graduate students in physical and inorganic chemistry as well as for more experienced investigators who are not already familiar with the subject. Much of it can also be understood by advanced undergraduate students.

In all important respects, the field of electron-transfer reactions has developed only since World War II. A clearer understanding of the general mechanisms has been established only in recent years. Before this time there were almost (well, not quite!) as many mechanisms as there were reactions. It is doubtful that a tentative discussion of the many mechanisms would at that time have been of more than transitory interest. In other respects, it must be admitted that the theory must be further refined in all reactions, and particularly in "non-adiabatic" reactions, before theory can enjoy our full confidence.

We have chosen to discuss a limited number of experimental studies in considerable detail so as to provide a basis for the analysis of other experimental data and to present theoretical results rather than derivations. Despite the constant flow of contributions to the body of factual knowledge concerning electron-transfer reactions and to the theory, this book should provide a general introduction to the subject for some time to come.

The preparation of this volume has been supported by several public agencies and one private foundation. Warren Reynolds gratefully acknowledges support from the United States Atomic Energy Commission. Rufus Lumry wishes to acknowledge support from the Louis and Maude Hill Family Foundation of St. Paul, Minnesota, and from the National Science Foundation. We both wish to thank Rudolph A. Marcus, Norman Sutin, and Svend E. Nielsen for their considerable assistance.

WARREN L. REYNOLDS
RUFUS W. LUMRY

Minneapolis, Minnesota
June, 1966

iii

Contents

MECHANISMS

OF

ELECTRON TRANSFER

I

Electron-Transfer Reactions

INTRODUCTION

Oxidation-reduction reactions are of importance in nature and technology. Their study has been a major activity in chemistry since the end of the war. This study has been so successful that the qualitative understanding of some reaction mechanisms is very nearly complete. Indeed, even the progress of the quantitative theories has been most satisfying. It is the purpose of this book to describe some experimental and theoretical highlights of that class of oxidation-reduction reactions in which one or more electrons migrate between reactants in the over-all reaction. A particularly simple reaction of this type is the transfer of one electron from tris-1,10-phenanthroline Fe(II) ion to a tris-1,10-phenanthroline Fe(III) ion as in

$$Fe(ph)_3^{+2} + {}^*Fe(ph)_3^{+3} \rightleftharpoons Fe(ph)_3^{+3} + {}^*Fe(ph)_3^{+2} \qquad (1\text{--}1)$$

No change in concentration of either Fe(II) or Fe(III) complex ion concentrations occurs, but there is a net transfer of an electron from the non-radioactive isotopes of iron in the $+2$ oxidation state to the radioactive isotope in the $+3$ oxidation state. Transfer is measured in this case by isotopic-tracer methods, but in many cases labelling is not essential since the methods of optical rotation, nuclear magnetic resonance, and electron paramagnetic resonance can measure the rate of reaction even if both iron ions are natural isotopes.

A slightly more complex example of an electron-transfer reaction, in which not only the electron but also a proton is transferred, is

$$Cr(H_2O)_6^{+2} + {}^*Cr(H_2O)_5(OH)^{+2} \rightleftharpoons Cr(H_2O)_5(OH)^{+2} + {}^*Cr(H_2O)_6^{+2} \qquad (1\text{--}2)$$

The small over-all free-energy changes associated with reactions (1–1) and (1–2) are due to isotopic mixing and to differences in zero-point energies. The free-energy changes are so small that they can be neglected in experiment and theory, so that to all intents and purposes the standard free-energy change, $\Delta F°$, is zero and the equilibrium constant is unity. We shall call these reactions *homonuclear* electron-transfer reactions to distinguish them

from the somewhat more complicated *heteronuclear* electron-transfer reactions, which involve electron transfer between reactants of different types.
The products of the latter are different from the reactants, so that there is a
net chemical change and $\Delta F°$ is not zero. A good example of a heteronuclear
electron-transfer reaction is

$$Fe^{+2}(aq) + IrCl_6^{-2} \to Fe^{+3}(aq) + IrCl_6^{-3} \qquad (1\text{--}3)$$

and like all such reactions, it can be studied with various simple methods,
depending on the different chemical properties of reactants and products.
Fortunately, methods are now available to study even the fastest of these
reactions, for example, those which are diffusion-controlled.

We shall be concerned only with homonuclear and heteronuclear electron-
transfer reactions and will neglect reactions, such as

$$IO_3^- + 5I^- + 6H^+ \to 3I_2 + 3H_2O \qquad (1\text{--}4)$$

in which oxidation states, coordination numbers of iodine atoms, and
hybridization of the iodine orbitals may change without any apparent
transfer of electrons between the iodine atoms. The iodine atoms in both
reactants and products have a complete octet of electrons in the usual
electronic structures assigned to the chemical species involved.*

Oxidation-reduction reactions of the type exemplified by Eq. 1–4 can be
written as a series of elementary unimolecular or bimolecular reaction steps,
some of which are rate-limiting and some of which are not. Some elementary
steps involve changes in oxidation states, while others do not. In those
elementary steps in which oxidation-reduction occurs, the making and
breaking of bonds accomplishes the change in oxidation states without any
necessity for net electron transfer. Such steps are those of conventional
chemical reactions in which strong interaction occurs between the reactants
with new chemical bonds of reasonable strength being formed between
reactants in the activated complex. Reactions of this sort are *adiabatic* in
the Ehrenfest sense. The entire reacting system always remains in a single
eigenstate of the total Schrödinger operator for the system, specifically, that
state which is usually the lowest for any nuclear configuration. Hence the

---

* The name "electron-exchange reactions" has been used to refer only to electron-
transfer reactions involving isotope exchange, i.e., our homonuclear type, whereas the
name "electron-transfer reactions" has been used in a more general sense (as we use it)
to include both of the types we designate as homonuclear and heteronuclear. Since
electron exchange can be said to occur between the reacting partners in the $Fe^{+2}(aq) +$
$Co^{+3}(aq) = Fe^{+3} + Co^{+2}(aq)$ reaction as well as in the $Fe^{+2}(aq) + *Fe^{+3}(aq) =$
$Fe^{+3}(aq) + *Fe^{+2}(aq)$ reaction or since electron transfer from $Fe^{+2}(aq)$ to $*Fe^{+3}(aq)$
as well as from $Fe^{+2}(aq)$ to $Co^{+3}(aq)$ can be said to occur, these names do not have
the virtue of distinguishing between their usages. In these usages, the name "electron-transfer reactions" could equally well be defined to mean our homonuclear
electron-transfer reactions and the name "electron-exchange reactions" to mean all
oxidation-reduction reactions in which an electron is transferred.

total reaction path is confined to a single electronic potential-energy surface. Most chemical reactions are of this type, and both the successes and failures of absolute reaction-rate theory in such cases are well known. It is an unfortunate fact that the quantitative success of any rate theories applied to such reactions is greatly restricted by the complexity of the quantum-mechanical calculation of potential energy. It is a fortunate aspect of many electron-transfer reactions that interactions between reactants in an activated complex are so weak that classical methods for calculation of potential energy are often adequate to the problem. This is particularly true for electron-transfer reactions between ions, and we may expect to have more success in calculations of absolute reaction-rate constants for such reactions than is generally obtained for other types of chemical reaction.

Interest in electron-transfer reactions has developed increasingly rapidly in the last fifteen years because of the availability of new experimental methods and instruments. Perhaps the most important such development has been the appearance of artificial isotopes, but new instrumentation for isotopic measurements and instrumentation for direct rate measurement in any time range have also served to open the field. In addition, there has been an increased interest in theoretical inorganic chemistry. In particular, we should mention the following major theoretical treatments of electron-transfer reactions:* Libby,[1] R. J. Marcus, Zwolinski and Eyring,[2,3] Weiss,[4] Laidler,[5] R. A. Marcus,[6,7,8,9,10,11] Halpern and Orgel,[12] Levich and Dogonadze,[13] McConnell,[14] Hush,[15] and Sacher and Laidler.[16] Many review articles on the subject have appeared. The recent reviews of Halpern,[17] Sutin,[18] and Marcus[19] will provide references to the important earlier reviews.

It is not out of place to point out for the benefit of those lacking firsthand familiarity with the field that the brilliant experimental work of Henry Taube and his associates forms a most important chapter in the recent history of electron-transfer reactions. References to this work are given throughout the text and should be consulted, not only for interesting details which were omitted here, but also to provide an unusually stimulating picture of science in action.

INNER-SPHERE ELECTRON-TRANSFER MECHANISMS

When two complex ion reactants share one or more ligands of their first coordination spheres in the activated complex for electron transfer, the activated complex is termed an *inner-sphere activated complex* and the mechanism an *inner-sphere mechanism*. In an investigation of the reduction of a number of Co(III) complex ions by $Cr^{+2}(aq)$, of which

$$Co(NH_3)_5Cl^{+2} + Cr^{+2}(aq) + 5H^+ \rightarrow Co^{+2}(aq) + CrCl^{+2} + 5NH_4^+ \quad (1-5)$$

* The numbers immediately following, and elsewhere, are keys to the References at the end of the book.

is typical, Taube, Myers, and Rich[20] found that a ligand was transferred from the substitution-inert inner sphere of the Co(III) complexes to the substitution-inert inner sphere of the product Cr(III) complex ion. Precisely because of the substitution-inert character of the Co(III) and Cr(III) complexes and the substitution-labile character of the $Cr^{+2}(aq)$ complex, the sharing of the transferred ligand between the inner coordination spheres of the two reactant ions in the activated complex was postulated. Such complexes were named "bridged transition states" or "bridged activated complexes"; but Halpern[17] later suggested that the name "inner-sphere activated complex" was more useful, since bridges may be formed between reactants in other kinds of activated complexes in which the reactants do not share ligands of their first coordination spheres. Halpern's term has persisted and we shall use it.

During the formation of the activated complex in reaction (1–5), the Cl^- ion attached to Co(III) apparently displaces a water ligand of $Cr(H_2O)_6^{+2}$ to give an activated complex of the formula $[(NH_3)_5Co \cdots Cl \cdots Cr(H_2O)_5^{+4}]^{\ddagger}$. After electron transfer, the labile $Co(II)—Cl^-$ bond readily breaks while the $Cr(III)—Cl^-$ bond is retained in the product. It is apparent that in order to demonstrate an inner-sphere activated complex in this simple way there must be a substitution-inert reactant, either Ox_1 or Red_2, and a substitution-inert product, either Ox_2 or Red_1, respectively. By substitution-inert reactant or product we mean a species inert toward substitution over a period of time long compared to that taken for the observation of the electron-transfer reaction between Ox_1 and Red_2. When this is not the case, inner-sphere mechanisms are more difficult to demonstrate. Their presence or absence is, of course, independent of the substitution-labile or substitution-inert character of the reactants and products. The reduction of $IrCl_6^{-2}$ by $Cr^{+2}(aq)$ is believed[21] to proceed through an inner-sphere complex though the products are $IrCl_6^{-3}$ and $Cr^{+3}(aq)$. Absence of chloride ion transfer probably means only that the $Ir(III)—Cl^-$ bond is less easily broken than the $Cr(III)—Cl^-$ bond.

The bridging group in an inner-sphere activated complex can perform several functions. The coulombic repulsion between ionic reactants of like charge is reduced if a bridging ion of opposite charge is placed between them. Similarily, coulombic repulsion is reduced if the bridging ligand is long so that electron transfer can occur without close approach of the charged reactants. Perhaps a much more important function of the bridge is to provide orbitals of proper symmetry to delocalize metal-ion electrons, thus providing a continuous pathway of good orbital overlap from metal ion to metal ion. The use of bridge orbitals, filled or unfilled, does not imply oxidation or reduction of the bridging group. Pictorially, the reductant loses an electron to the bridge as the bridge loses an electron to the oxidant in what has been termed a *"double-exchange"* mechanism.[12] On the other hand, bridge orbitals may not be involved in electron transfer, so that the bridge functions merely by bringing metal ions closer together for a *"direct exchange"* between overlapping metal orbitals.[12] Use of vacant bridge orbitals (usually of higher energy)[12,14] for electron delocalization and migration in a virtual process has been called *"superexchange."*[12*] Large electronic interactions between metal ions and bridge group can stabilize the activated complex and reduce the total free energy required for rearrangement of inner and outer coordination shells of the reactants prior to electron migration from one reactant to the other. Sutin[18] has pointed out that inner-sphere mechanisms appear to be preferred when rearrangement energies are large.

The bridging groups may also undergo temporary oxidation or reduction as a result of the metastability of an intermediate state in which both metal ions are either oxidized or reduced as a result of adding one or more electrons to the bridging ligand or removing one or more electrons from the bridging ligand, respectively. The lifetime of such states is long with respect to separate electron-transfer reactions. Frequently the lifetime is so long that chemical change in the bridge is possible, for example, a *cis-trans* isomerization. Halpern and Orgel[12] have suggested the name "chemical mechanism" for such reactions. A case in point is the reduction of penta-amminemaleato Co(III) by Cr^{+2}(aq) ion[22,23,24], which will be discussed in Chapter 3.

Inner-sphere mechanisms have also been called "group-transfer mechanisms," but the name is not useful. As we have seen, inner-sphere electron transfer can occur with or without transfer of the bridging group and there is no significant qualitative difference between the two kinds of process. The transfer of the group is incidental to electron transfer. It occurs, if at all, on decomposition of the products' activated complex as an accidental consequence of the relative substitution labilities of the bridging ligand in the metal complexes.

Relatively strong σ bonds are usually formed between the bridging ligand and the metal ions. Thus an N_3^- bridge will form σ bonds to both metal ions, using the lone-pair electrons of the end nitrogen atoms. Electron transfer appears to proceed through π molecular orbitals of the bridge and metal ions, as might be expected from symmetry. Generally, the π systems of bridge groups appear to be more efficient than σ systems in electron transfer. In the reaction between Fe^{+2}(aq) and FeN_3^{+2} there is evidence[25] for a dinuclear complex of the form Fe(II)—N_3^-—Fe(III), yet the over-all rate of the electron-transfer reaction is not unusually high. Under some conditions, the complex apparently dissociates with high probability before electron migration can occur. There is certainly a good π-orbital system for electron

---

* All these terms have been used in other or similar connections at earlier dates but Ref. 12 specifies their application to electron-transfer reactions.

transfer, however, so the relative slowness of the reaction is probably attributable to the high energy of the nuclear reorganizations required for electron transfer. Either the formation of the dinuclear complex or electron transfer within the dinuclear complex may be rate-limiting in this case, and there is evidence that both limits can be detected experimentally[25] (see Chapter 3).

In this particular example, the efficient pathway for electron transfer is a result of the fact that the migrating electron is in a t_{2g} orbital of the reductant and moves to a t_{2g} orbital of the oxidant. A low-lying π molecular orbital of the bridge has the proper symmetry to interact with both orbitals. On the other hand, if transfer involved an electron in an e_g orbital, symmetry restrictions would be expected to greatly reduce the transfer efficiency. For example, the transfer of an electron from an e_g orbital (really an antibonding σ molecular orbital) of Cr(II) to an e_g orbital of Cr(III) in Cr(III)—N_3^-—Cr(II) would be expected to be slow, since the antibonding σ orbital will mix poorly with the bridge π system. However, symmetry restrictions may be un-important, since the measured entropy of activation is the same for both the high-temperature Fe(II) + FeN_3^{+2} exchange and the corresponding Cr(II) + CrN_3^{+2} exchange (see pages 38 and 71 in Chapter 3).

FRANCK-CONDON RESTRICTIONS

Nearly all the special properties of electron-transfer reactions are due to Franck-Condon restrictions. These are familiar from the study of absorption and emission of light by molecules, and it takes but a moment of thought to conclude that the analogy between such light processes and electron-transfer reactions is very close. The "electron-jump" process involving a net transfer of an electron from an orbital belonging essentially to one metal to an orbital belonging essentially to the other metal occurs in a time short ($\sim 10^{-15}$ sec.) compared to that required for nuclear position change. Since nuclear motion, except for hydrogen, occurs in about 10^{-13} sec., or longer, nuclear position changes are one hundred times slower than the change in electronic state which occurs with the electron jump. The Franck-Condon principle requires that the nuclear positions remain essentially unchanged during the electronic transition. We must, therefore, consider two forms of the activated complex, both with the same spatial arrangement of nuclei, but one having the electronic description of the reactants and the other having the electronic description of the products. These complexes are designated the *reactants' activated complex* and the *products' activated complex*, respectively.

There are two major consequences of the Franck-Condon principle for electron-transfer reactions. The first is that the total energy of the reactants' activated complex must be identical, within the limits of the uncertainty principle, with the energy of the products' activated complex. That is, the energy of the activated complex as described by nuclear coordinates must be twofold degenerate, and degenerate in a special way that places the migrating electron on one reactant before transfer and on the other after transfer. If there were no degeneracy requirement, it would be possible to find a reaction in which the energy was higher in the products' activated complex than in the reactants' activated complex and one could then devise, at least in principle, ways in which to capture this energy to defeat the first law of thermodynamics. This situation and the resulting requirement for electronic degeneracy was recognized by Libby.[1] Marcus has generalized the restriction for a macroscopic reacting system of many activated complexes to take account of the second law of thermodynamics and to set a more stringent condition, since he finds that in such a system the free energy must remain constant during the electron-transfer act. In principle, neither the restriction on energy change nor that on free-energy change is rigorous, since the electronic system of the reactants is coupled to the electrons of environmental molecules and moves in phase with them. Hence it is possible that energy can be transferred into or out of the activated complex by this coupling during the electron-jump act. In practice, the coupling is too weak to make this a practical mechanism for energy transfer and we can consider the degeneracy requirement to be a strict one. Since nuclear motions are too slow for energy migration via atomic collisions, vibrational coupling, etc., the activated complex is essentially isolated, in the thermodynamic sense, during electron transfer.

In a homonuclear reaction, it is possible to reach the required activated complex by distorting ligands and solvent so as to form a symmetrical structure that undergoes no change on electron migration other than a rotation by 180°. Other structures will satisfy the free-energy restriction, but it is probable that the symmetrical structure is by far the most important single activated complex.

In a heteronuclear reaction, the electronic energy of the reactants' activated complex will usually be different from that of the products' activated complex. The energy-balance condition must then be achieved by reorganization of ligand shells and solvent in such a way that the vibrational energy of the ligand and solvent shells in the reactants' activated complex is different from the vibrational energy of the ligand and solvent shells in the products' activated complex by just the discrepancy in the electronic energy balance. That such conditions of balance exist is, of course, due to the fact that some vibrational-force constants depend on the position of the migrating electron and thus change on electron transfer. The reorganization of solvent and ligands, which has already been mentioned several times, is necessary to form activated complexes having the required condition of degeneracy. It is the major requirement for electron transfer between weakly interacting reactants and, together with coulombic-, translational-, and rotational-energy changes for placing the reactants at the interreactant distance of the activated complex, determines the free energy of activation.

In a homonuclear reaction, reorganization of solvent and ligands to form the reactants' activated complex involves equivalent distortions of the reactants. Thus in the $Fe^{+2}(aq) + Fe^{+3}(aq)$ reaction, the ligands about the ferrous ion are compressed while those about the ferric ion are expanded until the ligand–metal ion distance is the same in both. The geometry of the activated complex is between that of the normal reactants and normal products. If the electrostatic free energy of the reactants' activated complex is not too different from that of the products' activated complex, a similar situation exists for a heteronuclear reaction with a small over-all free-energy change. However, if there is a large negative over-all free-energy change in a heteronuclear reaction, the geometry of the activated complex is very similar to the normal geometry of the reactants. Little distortion is required to form the reactants' activated complex which, as a result, has little vibrational excitation. However, the products' activated complex in such reactions is highly distorted relative to the products' equilibrium geometry and thus has high vibrational excitation. Just the contrary situation exists if the over-all free-energy change is a large positive quantity. Hence, so far as the contribution to the free energy of activation from ligand and solvent reorganization is concerned, the experimental free energy of activation will tend to be low for negative over-all free-energy changes and high for positive over-all free-energy changes. This, of course, is what would be expected in any event. Hard-and-fast rules about the relationship between measured activation energy and reorganization of solvent and ligands in the activated complex are not possible, however, since the coulombic contribution to the experimental free energy of activation is highly variable and must be subtracted before analysis is possible.

If the over-all free-energy change is very large and negative, and suitable excited electronic states exist for products the products may be formed in excited electronic states. Similarly, for a large positive free-energy change, the mechanism may involve electronic promotion of reactants during the formation of the activated complex. Generally speaking, excited electronic states do not lie low enough to favor such mechanisms, but they undoubtedly occur in some instances. A variant on this mechanism concerns the change in spin quantum number if restricted by selection rules. The preferred process in such a case may involve change of the spin quantum number during the formation of the activated complex. The excited spin state would be such that the spin-selection rule did not restrict electron transfer.

The second consequence of the Franck-Condon principle is that no significant amount of angular momentum can be transferred into or out of the activated complex during electron transfer. Hence the angular momentum of the activated complex must be constant, and there will be selection rules which must be obeyed. The selection rules of importance will apply to electronic angular momentum and, in complex molecules, the only one likely to be of importance is that restricting changes of spin angular momentum. If spin-orbital coupling within reactants is strong, even this restriction will be unimportant. Since only orbitals of the same symmetry can be mixed by effective perturbations in electron-transfer reactions, there is at least in principle a set of symmetry-based selection rules due to conservation of orbital angular momentum that limits the magnitude of the electronic coupling between reactants' orbitals or among reactants and bridge orbitals responsible for electron migration. Just how important these are remains to be seen. Only small interactions are necessary; in complex molecules, orbital symmetry as it is involved here is not a rigorous concept since electron-nuclei momentum interchange is a rapid process. Certainly there will be some restrictions due to this cause, but the interaction need only be larger than some small minimum amount generally achievable even with symmetry restrictions.

The existence of an activated complex with the appropriate free-energy balance condition does not guarantee that the electron-transfer reaction will occur. There must be sufficient electronic coupling between reactants to allow transfer during the lifetime of the activated complex, which usually can be set at about 10^{-13} sec. or slightly less. We shall see in Chapter 5 that the probability of electron transfer can be estimated by procedures identical with those used to calculate the probabilities of light absorption or emission, except that the interaction operator is not the dipole moment operator. There are factors for vibrational overlap as well as electronic interaction just as in photoprocesses.

Since the electron-transfer process is formally identical with the interaction of any two electronic potential-energy surfaces, the full literature on crossing processes at the intersection or close approach of two such surfaces is applicable to electron-transfer acts. Thus much theoretical information relevant to electron-transfer processes already exists. However, to this time no specific analysis of these processes in terms of the existing literature on crossing processes has appeared.

ADIABATIC AND NON-ADIABATIC PROCESSES

It has proved useful, in discussing electron-transfer reactions, to take some liberties with Ehrenfest's definitions of adiabatic and non-adiabatic (diabatic) processes. "Strongly" adiabatic processes are those of ordinary chemical reactions, in which chemical bonds are broken and remade. The electronic interaction energy between the reactants makes a significant contribution to the energy of the activated complex.   The interaction energy is a resonance energy and will usually be referred to as such in this book.   Most electron-transfer reactions do not appear to be of the strongly adiabatic type even when reactants are bridged by sharing a common inner-shell ligand. There is a resonance energy in such reactions, but it is small enough to be ignored in calculating the free energy of activation (at least at the present time).   About 1 kilocalorie appears to be a maximum estimate at present for the resonance energy in many electron-transfer reactions.   The electron-transfer probability depends on the resonance energy, and 1 kilocalorie of resonance energy is more than sufficient to establish a unit probability for this transfer.   It is also large enough to justify calling such reactions adiabatic.   Thus, somewhat arbitrarily, we shall distinguish as *adiabatic* those reactions in which the electron-transfer probability is so large that there is essentially unit probability of transfer during the lifetime of the activated complex.   Such reactions occur without change of electronic quantum number, that is, they take place on a single electronic potential-energy surface.

When the resonance energy of the activated complex is very much smaller than 1 kilocalorie, the electron-transfer probability is small.   Very few of the reactant systems experience electron transfer as they pass through the configuration of the activated complex.   Such reactions correspond more closely to the non-adiabatic case of Ehrenfest and we shall call them *non-adiabatic* reactions, that is, we shall call reactions with a small electron-transfer probability (small transmission coefficient) non-adiabatic reactions. At intermediate resonance energies the distinction is not always sharp but, practically, the designations have significance insofar as the theoretical treatments to be applied can ignore the transfer-probability problem for adiabatic reactions but cannot for non-adiabatic reactions.   It is worthwhile noting that the electronic interaction between reactants approaching to van der Waals distances, even if both reactants have like charge, is probably sufficient in most cases for efficient electron transfer during the lifetime of the activated complex.   Insofar as this is true, most electron-transfer reactions are of the adiabatic type.

It is also to be noted that for both adiabatic and non-adiabatic electron-transfer reactions the electronic energy is not so high in either the reactants' or the products' activated complex as the classical potential-energy barrier to electron migration between these complexes.   Hence electron migration is a quantum-mechanical *tunnelling* process in which the electron "passes through" the potential barrier rather than over it.   This situation has led some theoreticians to attempt calculations of the transmission coefficient in terms of the penetration of the potential barrier by a plane wave.   We shall discuss these attempts; but it is now clear that this is not the best approach to the problem.   Conventional time-dependent perturbation methods are more realistic and easier to handle. They are also more easily incorporated within the framework of modern quantum-mechanical calculations. Most recent theoretical work has emphasized the latter approach, though the problem has not yet been treated in a very profound fashion even with this approach.

OUTER-SPHERE ELECTRON-TRANSFER MECHANISMS

Outer-sphere activated complexes are formed when the inner coordination shells of the reactant complex ions are left intact as to the number and kind of ligands present. Distortion (expansion, compression, or asymmetric rearrangement) of the inner coordination shells may occur and usually does. However, there is no sharing of common ligands between the inner coordination spheres. The occurrence of electron transfer through outer-sphere activated complexes is easily established when rapid electron transfer occurs between substitution-inert complex ions. In other cases, it is often extremely difficult to decide between inner-sphere and outer-sphere mechanisms.

Distortion of inner shells and rearrangement of outer solvation shells is necessary to satisfy the Franck-Condon restrictions, but when the required distortion of the inner shell is not greater than occurs in the ground-state vibrations of the reactants, no inner-shell rearrangement energy is necessary. Then only solvent reorganization need be considered. The homonuclear reaction between hexacyanoferrate(II) and (III) ions,

$$Fe(CN)_6^{-4} + *Fe(CN)_6^{-3} \rightleftharpoons Fe(CN)_6^{-3} + *Fe(CN)_6^{-4} \tag{1-6}$$

was once thought to be an example in which no inner-sphere reorganization was necessary as judged from the small activation energy for isotope exchange. Recent investigation has shown that the reaction as written may not represent the rate-determining electron-transfer step but that the exchange is catalyzed by cations for all cations investigated thus far. The catalyzed reactions are indeed fast and have a small activation energy. It is not clear whether the uncatalyzed reaction is (1) an inner-sphere or an outer-sphere reaction (although the fact that both reactants are substitution-inert strongly favors the latter), (2) slow because of ligand rearrangement energies or (3) slow because of a poor transmission coefficient. Reaction (1-6) has been considered the prototype of a well-understood electron-transfer reaction for many years, but it is now apparent that our understanding of some of these reactions is far from complete and that many current truths may turn out to be future fallacies. For this reason we have chosen to investigate a few well-studied electron-transfer reactions in detail in Chapter 3 rather than simply to provide tables of current conclusions about mechanism for the many electron-transfer reactions which have been studied. Fortunately, there are some reactions for which the picture is reasonably complete, but the number is not great.

Outer-sphere activated complexes may also be bridged. Reaction (1–6) occurs rapidly only when cations appear in the activated complex;[26] since the complex cyanides are substitution-inert, the cations probably play a bridging role between the inner coordination shells. The matter is discussed in more detail in Chapter 3. Other examples in which bridges are apparently formed in outer-sphere activated complexes are[27,28]

$$\text{Naph} + \text{Na}^+\text{Naph}^- \rightleftharpoons [\text{Naph} \cdots \text{Na}^+ \cdots \text{Naph}^-]^{\ddagger} \rightleftharpoons \text{Na}^+\text{Naph}^- + \text{Naph} \tag{1-7}$$

$$\text{dipy} + \text{K}^+\text{dipy}^- \rightleftharpoons [\text{dipy} \cdots \text{K}^+ \cdots \text{dipy}^-]^{\ddagger} \rightleftharpoons \text{K}^+\text{dipy}^- + \text{dipy} \tag{1-8}$$

in which Naph and Naph$^-$ are the naphthalene molecule and negative ion, respectively, and dipy and dipy$^-$ are the α,α'-dipyridyl molecule and negative ion, respectively. Na$^+$Naph$^-$ and K$^+$dipy$^-$ are undissociated complexes or ion pairs. The reactions are very fast. Consequently, it is probable that the reactants' geometry in the activated complex is nearly identical with that of the separated reactants. Bridging groups can perform the same functions in outer-sphere mechanisms as in inner-sphere mechanisms.

In non-bridged outer-sphere activated complexes, it is generally assumed that the adjusted inner coordination shells of the reactants are in van der Waals contact. This may not always be the case; there may be examples in which solvent remains between the reactants. However, unless such solvent provides an especially good pathway for electron migration, it is certainly reasonable to expect a much higher probability for electron transfer at van der Waals distances than at greater distances. The orbital overlap appears quite sufficient for efficient electron transfer in reactions such as

$$\text{Naph} + \text{Naph}^- \rightleftharpoons \{(\text{Naph} \cdots \text{Naph})^-\}^{\ddagger} \rightleftharpoons \text{Naph}^- + \text{Naph} \tag{1-9}$$

and it is quite clear from current knowledge of such orbitals that this can be the case only if the reactants approach at least as close as their van der Waals radii allow.

The hydrogen-atom transfer mechanism in electron-transfer reactions has usually been discussed as a separate mechanism different from those thus far presented. However, it can be discussed just as easily as a hydrogen-bonded outer-sphere mechanism. For example, the reactions

$$\text{Fe(H}_2\text{O)}_6^{+2} + {}^*\text{Fe(H}_2\text{O)}_5(\text{OH})^{+2} \rightleftharpoons$$

$$\left\{\left[\underset{\quad}{(\text{H}_2\text{O})_5\text{Fe}-\overset{\overset{\displaystyle H}{|}}{\text{O}} \cdots H \cdots \overset{\overset{\displaystyle H}{|}}{\text{O}}-{}^*\text{Fe(H}_2\text{O})_5}\right]^{+4}\right\}^{\ddagger} \rightleftharpoons$$

$$\text{Fe(H}_2\text{O)}_5(\text{OH})^{+2} + {}^*\text{Fe(H}_2\text{O)}_6^{+2} \tag{1-10}$$

$$Fe(H_2O)_6^{+2} + {}^*Fe(H_2O)_6^{+3} \rightleftharpoons$$

$$\left\{\left[\begin{array}{c} \overset{H\quad H}{(H_2O)_5Fe-O \times O-{}^*Fe(H_2O)_5} \\ H\quad H \end{array}\right]^{+5}\right\}^{\ddagger} \rightleftharpoons \qquad (1\text{--}11)$$

$$Fe(H_2O)_6^{+3} + {}^*Fe(H_2O)_6^{+2}$$

can be formulated in this way, though their actual electron-transfer mechanisms are unknown. The hydrogen bonds need not be linear. Net hydrogen transfer can occur in reaction (1–10), but such transfer is not required by the principle of detailed balancing (Chapter 3). It is nevertheless highly probable, as a consequence of the different charges on the products. However, this reaction need not be a conventional direct electron-transfer reaction, since the hydrogen atom migrating from left to right produces the same result. On paper, the mechanisms are distinctly different, but experimental methods for distinguishing between the two have not yet been developed. Such hydrogen-atom transfers do not look unreasonable from an energetic point of view and may be important.

In reaction (1–11), two hydrogen bonds have been pictured; Horne[29] has gone further in postulating four hydrogen bonds involving all four of the protons on two water molecules, one from each reactant. He has done so to provide a symmetrical activated complex; but no rigorous requirement for symmetry actually exists. The activated complex shown in Eq. 1–11 could dissociate along either of the dashed lines to give products identical to reactants. If only one hydrogen bond were formed in the activated complex, as in

$$Fe(H_2O)_6^{+2} + {}^*Fe(H_2O)_6^{+3} \rightleftharpoons$$

$$\left\{\left[\begin{array}{c} \overset{H\qquad\ H}{(H_2O)_5Fe-O \cdots H \cdots O-{}^*Fe(H_2O)_5} \\ H \end{array}\right]^{+5}\right\}^{\ddagger} \rightleftharpoons \qquad (1\text{--}12)$$

$$Fe(H_2O)_6^{+3} + {}^*Fe(H_2O)_6^{+2}$$

dissociation along the dashed line would again lead to products identical with reactants. Note, however, that the electron-transfer path shown in this equation requires that the configurations

$$\left[\begin{array}{c} \overset{H\qquad\ H}{(H_2O)_5Fe^{II}-O \cdots H \cdots O-Fe^{III}(H_2O)_5} \\ H \end{array}\right]^{+5}$$

and

$$\left[(H_2O)_5Fe^{III}\!-\!\overset{\underset{\displaystyle H}{|}}{O} \cdots H \cdots \overset{\displaystyle \overset{H}{\diagdown}}{\underset{\displaystyle \diagup}{O}}\!-\!Fe^{II}(H_2O)_5 \right]^{+5}$$

have equal energies within the range allowed by the uncertainty principle (the water molecules not involved in the hydrogen bond have been correctly rearranged to satisfy Franck-Condon restrictions). It is not clear that this can be the case.

To what extent the bridge bonds to the hydrogen atom might participate in the transfer of the t_{2g} electron from Fe(II) to Fe(III) is not clear. Perhaps they would merely stabilize the activated complex for "direct exchange," rather than participating in "double" or "superexchange."

Inner-sphere mechanisms are more complicated than outer-sphere mechanisms because at least one inner-shell ligand must be displaced from one reactant or at least one ligand shell must be rearranged to allow for an increase in coordination number. They are both difficult to assess theoretically. There are few experimental data directly applicable to ligand substitution reactions between two complex ions, so few in fact that we shall not be able to say much more about the ligand-displacement reactions in this monograph. In any event, in addition to the possibilities that the over-all reaction can be limited either by diffusion or by the ligand and solvent rearrangement, there also exists the very real possibility that inner-sphere reactions have their rates determined by the ligand-displacement step, as has already been discussed for the $Fe^{+2}(aq) + FeN_3^{+2}$ reaction. Since the ligand-displacement step is absent in outer-sphere reactions, the fastest electron-transfer reactions are probably of that type. There are some very fast reactions. For example, the reaction between tris-(4,7-dimethyl-1,10-phenanthroline)Fe(II) and hexachloroiridate(IV) has a forward-rate constant of 1×10^9 liter/mole sec. at 10°C and a backward-rate constant as high as 4×10^9 liter/mole sec. at 30°C.[30] Both forward and backward reactions are very close to the diffusion limit and it is probable that diffusion rates and electron-transfer rates are about equal. Ligand and solvent rearrangement energies are very small, and the reaction is undoubtedly of the outer-sphere type. The rate constant for electron exchange between naphthalene and naphthalene negative ion in the presence of Na^+ is of the same order of magnitude[31] and, since it involves one reactant with net zero charge (in contrast to the oppositely charged reactants of the previous reaction), is probably diffusion-controlled. These reactions are a bit unusual; most electron-transfer reactions are slow relative to the diffusion times even when reactants have large charges of the same sign and thus relatively low diffusion rates.

Although resonance interaction necessary for electron transfer generally can be expected to be smaller in outer-sphere activated complexes than in inner-sphere activated complexes, available experimental evidence does not indicate a preference for non-adiabatic mechanisms in outer-sphere reactions.

SPECIAL MECHANISMS

If reaction (1–10) occurs by way of hydrogen-atom transfer rather than direct electron transfer, it and similar reactions might be bridged not by a single hydrogen bond but by a short chain of water molecules, all hydrogen-bonded to each other to provide a path for hydrogen-atom transfer not unlike the Grotthuss conduction mechanism. Such a mechanism was first proposed by Dodson and Davidson[32] and was later discussed by Reynolds and Lumry,[33] who pointed out that long bridges of this sort would cut down electrostatic repulsion and could explain isotopic effects. It now appears that neither argument is valid. However in terms of current ideas about the covalent character of hydrogen bonds, hydrogen-atom migration by this path is certainly reasonable if imperfect phase relations along the path do not require activated complexes of high energy. This last condition may set the upper limit to bridge length. The process may be of some importance in biological systems, though there is no evidence requiring it in such systems. It may occur in ice, and has been discussed for such conditions by Horne and Axelrod,[34] who point out a number of similarities between proton conductance and some electron-transfer reactions. Though it will not be further dealt with in this monograph, the last word remains to be said, and the mechanism will certainly continue to appear in the literature of electron-transfer processes.

In solvents such as liquid ammonia, the solvation of electrons stabilizes them and makes electron migration over long distances easy and important. Hydration of electrons also occurs, but at high energy, so that thermal electron-transfer reactions via hydrated electrons are quite improbable except, of course, in photoreactions.

2

Metal Ion–Solvent Bond Energies

Since reactions of the type under discussion have been studied chiefly as reactions between ionic species in solution, it is desirable to consider very briefly some aspects of the nature and thermodynamics of ions in solution. The volume of literature concerning these subjects has become so great that only a few of the more pertinent references[35-40] will be given; from these the reader can obtain a rather complete list of literature references.

SOLVATION ENERGY

In solution, an ion is solvated. Much of the energy required to ionize a metal atom (ion) is furnished by the interaction between the resulting ion and the solvent. The interaction between ion and solvent usually produces changes in the solvent structure as compared to pure solvent at the same temperature and pressure, the actual changes depending upon the size of the ion and upon the strength of the electrostatic field surrounding the ion. The ion-solvent interaction energy and the change of the solvent-solvent inter-action energy contribute to the solvation energy of the ion. The solvation energy is the energy change associated with the solvation reaction

$$\text{Ion(g)} + \text{pure solvent} \rightarrow \text{Ion (solution phase)} \qquad \Delta X_1 \qquad (2\text{-}1)$$

The change of a thermodynamic function, ΔX_1, accompanying reaction (2–1) may be calculated[41] with the aid of Eqs. 2–2, 2–3, 2–4, and 2–5.

$$\tfrac{1}{2}H_2(g) = H^+(g) + e^-(g) \qquad \Delta X_2 \quad (2\text{-}2)$$

$$M(\text{standard state}) = M^z(g) + ze^-(g) \qquad \Delta X_3 \quad (2\text{-}3)$$

$$M(\text{standard state}) + zH^+(aq) = M^z(aq) + (z/2)H_2(g) \qquad \Delta X_4 \quad (2\text{-}4)$$

$$M^z(g) + zH^+(aq) = M^z(aq) + zH^+(g) \qquad \Delta X_5 \quad (2\text{-}5)$$

For reaction (2–5), ΔX_5 is simply the difference between $\Delta X_1(M^z)$ for the solvation of the metal ion M^z and $z\,\Delta X_1(H^+)$ for the solvation of zH^+ ions (if H^+ is used as a reference), so that

$$\Delta X_5 = \Delta X_1(M^z) - z\,\Delta X_1(H^+) \qquad (2\text{-}6)$$

Also ΔX_5 can be expressed as

$$\Delta X_5 = \Delta X_4 - \Delta X_3 + z\,\Delta X_2 \qquad (2\text{-}7)$$

and hence ΔX_5 may be calculated for a number of ions. When a value for ΔX_5 is substituted in Eq. 2–6, $\Delta X_1(M^z)$ for the solvation reaction of the metal ion may be obtained in terms of the value of the corresponding change for H^+. If a reasonably good value for $\Delta X_1(H^+)$ can be selected by some means, the value of $\Delta X_1(M^z)$ can then be obtained.

SOLVATION NUMBER

The total number of solvent molecules contributing a measurable amount to the solvation energy, i.e., the total solvation number, of a specific ion is unknown. With the exception of a few ions, most notably Cr^{+3} (see Ref. 42), not even the primary hydration numbers of most ions are known. The primary hydration number, defined to be the number of water ligands in the first coordination sphere, is usually assumed to be the same as the most common coordination number of the ion.

HYDRATION ENERGIES OF Fe^{+3}

For a much more complete discussion of metal ion–solvent bond energies, the reader should consult Ref. 43 (especially Chapter 2). Inasmuch as many of the homonuclear electron-transfer reactions studied have been between complexes containing water molecules in the first coordination shell, it is of special interest to calculate the average bond energies of the metal ion—OH_2 bond in these complexes. For example, this bond energy at 25°C for Fe^{+3} may be calculated from the following set of reactions:[43]

	ΔH (kcal.)	ΔF (kcal.)	
$6\ H_2O(l) = 6\ H_2O(g)$	63	12	$(2\text{-}8)$
$Fe^{+3}(g) + 6\ H_2O(g) = Fe(H_2O)_6^{+3}(g)$	$-6\overline{BE}_H$	$-6(\overline{BE})_F$	$(2\text{-}9)$
$Fe(H_2O)_6^{+3}(g) = Fe(H_2O)_6^{+3}(aq)$	-438	-430	$(2\text{-}10)$
Sum: $Fe^{+3}(g) + 6\ H_2O(l) = Fe(H_2O)_6^{+3}(aq)$	-1050	-1010	$(2\text{-}11)$

in which $\overline{BE}$ is the average bond energy of an Fe^{+3}—OH_2 bond in $Fe(H_2O)_6^{+3}(g)$ at the Fe^{+3}—OH_2 distance which prevails in $Fe(H_2O)_6^{+3}(aq)$.

The value of -1050 kcal./mole for ΔH of the hydration reaction (2–11) of

Fe^{+3} was calculated from Eqs. 2–7 and 2–6, using appropriate thermodynamic data[44] for Eq. 2–7 and a value of -263 kcal./mole[45] for the enthalpy of hydration of H^+ in Eq. 2–6.

The value of -1010 kcal./mole for ΔF of reaction (2–11) was obtained in a similar manner, using a value of -5.5 cal./deg. mole[35] for the entropy of $H^+(aq)$.

BORN CHARGING

The values of ΔH and ΔF for reaction (2–10) were obtained from a Born charging process.[43] The uncharged species $Fe(H_2O)_6$ is considered to be spherical and to have a conducting surface (recall spreading of charge over the surface by the electroneutrality principle). The difference between the work, ΔF_ε, of charging the ion in a solvent of dielectric constant ε and the work, ΔF_v, of charging in a vacuum is called the Born free energy of hydration, ΔF_B, and is given by

$$\Delta F_B = \int_{q=0}^{q=ze} \frac{q\,dq}{\varepsilon r} - \int_{q=0}^{q=ze} \frac{q\,dq}{r} = -\frac{z^2 e^2}{2r}\left(1 - \frac{1}{\varepsilon}\right) \qquad (2\text{–}12)$$

where $z =$ valence of ion, $e =$ absolute value of the electronic charge, and $r =$ the radius of $Fe(H_2O)_6^{+3}$, i.e., the ionic radius of Fe^{+3} plus the diameter of a water molecule.

The corresponding entropy and enthalpy changes were obtained from

$$\Delta S_B = -\left(\frac{\partial(\Delta F_B)}{\partial T}\right)_P = \frac{z^2 e^2}{2r\varepsilon^2}\left(\frac{\partial\varepsilon}{\partial T}\right)_P \qquad (2\text{–}13)$$

$$\Delta H_B = \Delta F_B + T\,\Delta S_B \qquad (2\text{–}14)$$

with $(\partial E/\partial T)_P$ taken to be -0.36/deg. at $25°C$.[43]

"EXPERIMENTAL" BOND ENERGIES

With values of ΔH and ΔF for reactions (2–8), (2–10), and (2–11), the so-called "experimental" average bond energies in Eq. 2–9 may be calculated. In this example the average values of ΔH and ΔF per bond are 113 kcal. and 99 kcal., respectively. For reaction (2–9), ΔS per bond is -45 cal./deg. mole, and is the value to be expected if a water molecule from the gas phase becomes tightly bonded to the metal ion. The good agreement between calculated and expected values of ΔS must be regarded as fortuitous in view of the approximate nature of some of the quantities involved in the calculation. (It is to be noted that the ΔH and ΔF calculated above are not the same as the bond energy at $0°K$.)

EXTENT OF DIELECTRIC SATURATION

The "experimental" values of the Fe^{+3}—OH_2 bond energies found in the preceding paragraph would be incorrect if dielectric saturation were to extend into the second hydration sphere of Fe^{+3}, because the Born charging energies would be incorrect. If dielectric saturation occurred to some extent in the second hydration sphere, then ε in Eq. 2–12 would be less than the macroscopic value and ΔF_B and ΔH_B would be algebraically greater than the values given in Eq. 2–10. This would cause the calculated bond energies in reaction (2–9) to become more positive than the values of 112 and 99 kcal. reported above; the values of 112 and 99 kcal. would be lower limits. Using these lower limits for first-coordination-sphere bond energies, upper limits for second-coordination-sphere bond energies can be obtained as follows. Eight molecules in the second coordination sphere are placed against the eight octahedral faces of $Fe(H_2O)_6^{+3}$ so that each of the eight water molecules— assumed to be hard spheres—just touches three water molecules forming an octahedron face in the first coordination shell. The resulting $Fe(H_2O)_{14}^{+3}$ ion is assumed spherical with a radius of 4.75 Å, the distance from the center of Fe^{+3} out through a diameter of a second-sphere H_2O molecule. Now $\Delta F_B =$ -314 kcal./mole, $\Delta H_B = -319$ kcal./mole, and the average bond energy is approximately 15 kcal./bond for each second-sphere molecule. This value for the bond energy between Fe^{+3} and a second-shell H_2O molecule is hardly sufficient to hold the ligand rigidly aligned, so that significant dielectric saturation in the second shell is unlikely.

CALCULATED BOND ENERGIES

Values of metal-ligand bond energies obtained from a series of reactions such as those for Fe^{+3} may be compared with bond energies calculated from the simple electrostatic theory of the coordinate bond. Basolo and Pearson[43] have tabulated the results of a number of such calculations. The expression for the potential energy of an octahedral complex composed of a central metal ion and six identical, electrically neutral ligand molecules is, in the gas phase,

$$U = \frac{-6q(\mu_0 + \mu_i)}{r^2} + \frac{6(1.19)(\mu_0 + \mu_i)^2}{r^3} + \frac{6\mu_i^2}{2\alpha} + \frac{6\beta}{r^9} \qquad (2\text{–}15)$$

where $q =$ charge on the central metal ion, $r =$ radius of the central metal ion plus the radius of a water molecule, $\mu_0 =$ magnitude of the permanent dipole moment of a ligand molecule, $\mu_i =$ magnitude of the induced dipole moment of a ligand molecule, $\alpha =$ the polarizability of the ligand, and β is a constant. The potential-energy zero is taken at infinite separation of the metal ion and all ligand molecules at rest in the gas phase. The first term on the right represents interaction between the central metal ion and the six polar ligands treated as point dipoles located at the center of the ligand molecule and pointing toward the central ion. The second term represents the mutual repulsion of the dipoles. The third term represents the energy used in forming the induced dipoles and the fourth represents the repulsion forces between the ligands and the central ion. The induced moment μ_i is given by

$$\mu_i = \alpha|\mathbf{F}| \tag{2-16}$$

where

$$\mathbf{F} = \frac{q}{r^2}\mathbf{K}_r - \frac{2.37(\mu_0 + \mu_i)}{r^3}\mathbf{K}_r \tag{2-17}$$

and $\mathbf{K}_r$ is a unit vector pointing from metal to ligand and lying along the line of centers. $\mathbf{F}$ is the field at the center of a ligand due to the ion and the other ligands of the first coordination sphere. Substituting for $\mathbf{F}$ in Eq. 2-16 and solving, it is found that the magnitude of the induced moment is given by

$$\mu_i = \frac{\alpha(qr - 2.37\mu_0)}{2.37\alpha + r^3} \tag{2-18}$$

Substitution of this expression for μ_i in Eq. 2–15 gives

$$U = 3\frac{2.37\mu_0^2 r - 2\mu_0 qr^2 - \alpha q^2}{r(2.37\alpha + r^3)} + \frac{6\beta}{r^9} \tag{2-19}$$

In calculating "theoretical" ΔH and ΔF values to compare with the "experimental" ΔH and ΔF values at 25°C of reaction (2–9), one should add to U a correction for the change of translational and rotational energies in the complexing reaction. The reactants at 25°C are first brought to rest, thus losing their translational and rotational energies. The complex ion is then formed from the reactants at rest, yielding the complex ion at rest and the energy U given by Eq. 2–19. The complex ion is given translational and rotational energies commensurate with the temperature of 25°C. The translational enthalpies will be taken as $\frac{5}{2}RT$ and the rotational enthalpies as $\frac{3}{2}RT$, except for $Fe^{+3}(g)$, which has negligible rotational enthalpy. The corrected value for the "theoretical" enthalpy change of reaction (2–9) is

$$\Delta H = U - 22.5RT \tag{2-20}$$

The "theoretical" free-energy change of reaction (2–9) is

$$\Delta F = -RT \ln Q + U \tag{2-21}$$

in which $Q = (Q_C N^6)/(Q_{Fe}Q_L^6)$, with Q_C, Q_{Fe}, and Q_L the partition functions of the complex ion, Fe^{+3}, and the ligand, respectively, and N is Avogadro's number. The partition functions are products of the translational and rotational partition functions for the polyatomic species; no vibrational partition function is included in this treatment.

The Born enthalpy and free-energy changes in Eq. 2–10 must be added to Eqs. 2–20 and 2–21, respectively, as in

$$\Delta H = U - 22.5RT - \frac{z^2 e^2}{2(r + r_{H_2O})}\left(1 - \frac{1}{\varepsilon}\right) + \frac{z^2 e^2 T}{2\varepsilon^2(r + r_{H_2O})}\left(\frac{\partial \varepsilon}{\partial T}\right)_P$$

$$(2\text{--}22)$$

$$\Delta F = -RT \ln Q + U - \frac{z^2 e^2}{2(r + r_{H_2O})}\left(1 - \frac{1}{\varepsilon}\right) \qquad (2\text{--}23)$$

Changes of translational and rotational energies should also be taken into account in transferring the complex ion from the gas phase to the solution phase at 25°C. However, these changes were neglected in calculating the "experimental" values and therefore, for the sake of consistency, they will also be neglected here. Equations 2–22 and 2–23 represent energies involved in taking reactants at 25°C in the gas phase, making a complex ion out of them, and transferring the complex ion to the solution phase. In these equations the radius of the complex ion being charged is $(r + r_{H_2O})$, where r is the center-to-center distance in Eq. 2–19 and r_{H_2O} is the radius of the ligand (1.38 Å for a water molecule).

In order to calculate a value for the repulsion parameter β in Eq. 2–19, the free-energy expression Eq. 2–23 must be minimized with respect to r, the derivative set equal to zero, and the resulting expression rearranged to give an equation for β. Sacher and Laidler[16] minimized an expression for enthalpy change; the difference is negligible. It will be assumed that the rotational partition functions are constants; in particular, the moment of inertia of the complex ion will be assumed to be that for the equilibrium distances in the complex ion. Minimization gives

$$\frac{6\beta}{r^{10}} = \frac{(2.37\mu_0^2 - 4\mu_0 qr)}{3r(2.37\alpha + r^3)} + \frac{(2\mu_0 qr^2 - 2.37\mu_0^2 r + \alpha q^2)(4r^3 + 2.37\alpha)}{3r^2(2.37\alpha + r^3)^2}$$

$$(2\text{--}24)$$

$$+ \frac{z^2 e^2}{18(r_c)^2}\left(1 - \frac{1}{\varepsilon}\right)$$

For $r = 2.05$ Å, $r_c = 2.05 + 1.38 = 3.43$ Å, $\mu_0 = 1.85$ debye, and $\alpha = 1.48 \times 10^{-24}$ cm.^{3}, it is found that the numerical value of β is 1.90×10^{-81}.

Using this value of β in Eq. 2–20, it is found that $|\Delta H|$ per bond for the gas-phase reaction at 25°C is 108 kcal./bond. In the evaluation of Eq. 2–21 the translational partition functions, $Q_t = (2\pi mkT)^{3/2}V/h^3$, and the rotational partition function, $Q_r = 8\pi^2(8\pi^3 A BC)^{1/2}(kT)^{3/2}/\sigma h^3$, were used. For the water molecule, A, B, C were set equal to 0.996×10^{-40}, 1.908×10^{-40} and

2.981×10^{-40} g. cm.^{2}, respectively, and $\sigma = 2$; for $Fe(H_2O)_6^{+3}$, $A = B = C = 502.8 \times 10^{-40}$ g. cm.^{2} and $\sigma = 24$. Hence it is readily calculated that $|\Delta F| = 97.5$ kcal./bond. These "theoretical" values of ΔH and ΔF are in deceptively good agreement with the "experimental" values. Because of its many crude approximations, the electrostatic model cannot be relied upon for the calculation of over-all enthalpies and free energies of reactions such as reaction (1–3). The reason for this is that ΔH and ΔF values calculated by Eqs. 2–22 and 2–23 are very large, and a 5-per cent error in them may be comparable to or larger than the over-all value being calculated.

However—and this is important—the electrostatic method can be used satisfactorily in many cases to calculate *changes in the enthalpy and free energy of an ion accompanying small changes in the coordination and solvation spheres*. This is because the errors in the values of $\Delta H(r_1)$ and $\Delta H(r_2)$ at the two extremes of the change will largely cancel in the difference $\Delta H(r_1) - \Delta H(r_2)$ since the errors will be in the same direction. Electrostatic calculations have been used by Marcus,[6] Hush,[15] Sutin,[18] and Sacher and Laidler[16] to obtain the contribution to the activation free energy from the rearrangement of ligand and solvent molecules; these calculations will be discussed at greater length when the theories of electron transfer are discussed. Although admittedly the electrostatic calculations are very crude, they will be much more successful in giving activation energies for the reactions under discussion than will Hamiltonian operators and wave functions for some time to come.

REARRANGEMENT ENERGIES

Changing ligand and solvent equilibrium configurations around reactant ions to a non-equilibrium configuration has been called variously "reorientation," "reorganization," and "rearrangement." We will call this process *rearrangement*, and denote the enthalpy and free energy changes by ΔH_{ar} and ΔF_{ar}. ΔH_{ar} and ΔF_{ar} for the change $r_1 \to r_2$ are given by

$$\Delta H_{ar} = \Delta H(r_2) - \Delta H(r_1) \qquad (2\text{–}25)$$

$$\Delta F_{ar} = \Delta F(r_2) - \Delta F(r_1) \qquad (2\text{–}26)$$

in which $\Delta H(r)$ and $\Delta F(r)$ are obtained from expressions analogous to those in Eqs. 2–22 and 2–23. In the differences, the term due to the change of translational and rotational energy in gas-phase reaction (2–9), namely, (constant) RT, will cancel from the ΔH expression, and the term

$$- RT \ln \left(N^6/Q_{Fe}Q_L \right)$$

will cancel from the ΔF expression. In particular, for $Fe(H_2O)_6^{+3}$, Eqs.

2–25 and 2–26 become

$$\Delta H_{ar} = U_2 - U_1 - \frac{z^2 e^2}{2}\left\{\left(1 - \frac{1}{\varepsilon}\right) - \frac{T}{\varepsilon^2}\left(\frac{\partial \varepsilon}{\partial T}\right)_P\right\}\left\{\frac{r_1 - r_2}{(r_1 + r_{H_2O})(r_2 + r_{H_2O})}\right\}$$

$$(2\text{-}27)$$

$$\Delta F_{ar} = \frac{3}{2}RT \ln\left(\frac{I(r_1)}{I(r_2)}\right) + U_2 - U_1 - \frac{z^2 e^2}{2}\left(1 - \frac{1}{\varepsilon}\right)\left\{\frac{r_1 - r_2}{(r_1 + r_{H_2O})(r_2 + r_{H_2O})}\right\}$$

$$(2\text{-}28)$$

where $I(r_1)$ and $I(r_2)$ are the moments of inertia for the complexes characterized by r_1 and by r_2, respectively. For displacements Δr, which are only a small fraction of the value of r_1, the first term for ΔF_{ar} will contribute less than 1 kcal., a negligible quantity.

LIGAND-FIELD STABILIZATION ENERGIES

Thus far, the contribution of the ligand-field stabilization energy (LFSE) to the bond energy of the metal-ligand bond has been neglected. In the case of $Fe(H_2O)_6^{+3}$ the LFSE is zero so that the neglect is justifiable but, in general, the value of U in Eq. 2–15, calculated from the electrostatic model, must be corrected for this additional stabilization energy. When a transition state in a reaction is formed from reactants, the accompanying change of LFSE should be taken into account. The change of LFSE arises from a change of metal ion–ligand distance or from a change in the number, symmetry, and kind of ligands surrounding a metal ion.

It is much beyond the scope of this monograph to discuss in any detailed fashion the calculation of LFSE. Our purpose is only to mention briefly how the term originates and how approximate estimates of LFSE may be made. The reader may consult other sources[43,46,47] for more detailed accounts and further references.

In a complex ion, LFSE arises from the splitting of a set of orbitals (which are degenerate in the free gaseous metal ion) by the electric fields associated with the ligands of the complex and from the distribution of electrons among these perturbed orbitals. Convenient examples in a discussion of ligand-field effects are the octahedral complex ions of the metals of the first transition period, in which all six ligands are alike. In the absence of an electric or magnetic field, the five $3d$ orbitals in these metal ions all have the same energy. However, when six ligands are placed in octahedral arrangement on the x, y, and z axes around the central metal ion, their electric fields in this particular symmetry split the $3d$ orbitals into two degenerate sets of orbitals. Three orbitals, the d_{xy}, d_{xz}, and d_{yz} orbitals of the metal ion, have an energy which is less than that for a spherical distribution of charge about the ion in the presence of the ligands because these three orbitals are directed away from the electron-dense regions of the ligands. Two orbitals, the d_{z^2} and $d_{x^2-y^2}$ orbitals of the metal ion, have an energy which is greater than that for a spherical distribution of charge about the ion in the presence of the ligands because these two orbitals are directed toward the electron-dense regions of the ligands. The energy difference between the upper and lower sets of $3d$ orbitals is defined as 10 Dq.

The first three electrons in the $3d$ orbitals of the metal ion go unpaired into the d_{xy}, d_{xz}, and d_{yz} orbitals. The stabilization energy is -4 Dq, neglecting electron-electron repulsion, for each electron to go into these orbitals. The LFSE for d^1, d^2, and d^3 complexes would be -4 Dq, -8 Dq, and -12 Dq, respectively. The fourth and fifth electrons may go into either the upper or lower set of orbitals, depending on the value of the energy spacing, 10 Dq, between the two sets, as compared to the energy required to pair electrons in the ions considered. If 10 Dq is less than the energy required for pairing, then the fourth and fifth electrons will go into the d_{z^2} and $d_{x^2-y^2}$ orbitals producing the "spin-free" or "outer-orbital" coordination complexes. The LFSE is decreased by 6 Dq for each electron which goes into the d_{z^2} and $d_{x^2-y^2}$ orbitals. Thus the spin-free complexes with d^4 and d^5 configurations have LFSE of -6 Dq and 0 Dq, respectively. The sixth, seventh, and eight electrons go into the d_{xy}, d_{xz}, and d_{yz} orbitals, giving spin-free complexes with LFSE of -4 Dq, -8 Dq, and -12 Dq, respectively. The ninth and tenth electrons go into the d_{z^2} and $d_{x^2-y^2}$ orbitals giving complexes with LFSE of -6 Dq and 0 Dq, respectively.

If 10 Dq is greater than the energy required for electron pairing, then the fourth, fifth, and sixth electrons will go into the d_{xy}, d_{xz}, and d_{yz} orbitals, producing the "spin-paired" or "inner-orbital" complexes. The LFSE of these three complexes are -16 Dq, -20 Dq, and -24 Dq, respectively, uncorrected for the spin-pairing energies. The spin-pairing energy is the coulombic and exchange energy required to take an electron from an upper d orbital and place it in a lower d orbital already occupied by one electron. The values[48] of this quantity per electron pair are approximately 67, 73, 50, and 64 kcal. for d^4, d^5, d^6, and d^7 divalent ions of the first transition series and 80, 86, 60, and 77 kcal. for d^4, d^5, d^6, and d^7 trivalent ions of the same series. Thus the LFSE of a Cr(II)(d^4) spin-paired octahedral complex would be decreased by 67 kcal./mole as the result of the formation of one electron pair and those of Mn(II) (d^5) and Fe(II) (d^6) would be decreased by 146 and 100 kcal./mole, respectively, for the formation of two electron pairs each. The seventh to tenth electrons, inclusive, would go into the d_{z^2} and $d_{x^2-y^2}$ orbitals, decreasing the LFSE from -24 Dq in d^6 to 0 Dq in d^{10}.

Complex ions of symmetries other than octahedral produce different d-orbital splitting. The most important cases are summarized in Ref. 43 (p. 55).

The value of Dq varies with the nature of the metal ion and the ligand and is best evaluated from the spectrum of the ion. (See Ref. 43, p. 44, for a tabulation of Dq values for divalent and trivalent aquo metal ions.)   However, approximate expressions have been derived for 10 Dq for ion and dipole ligands[49] and these may be used when necessary.

The total energy contribution from LFSE, plus that of electron pairing and electron correlation, should be taken into account in the calculation of enthalpies and entropies of activation.   In the case of the $Fe(H_2O)_6^{+2}$ + $Fe(H_2O)_6^{+3}$ exchange, if it proceeded by the outer-sphere activated complex mechanism with all iron-water distances equal to 2.13 Å, the change of LFSE for $Fe(H_2O)_6^{+2}$ stabilizes the transition state by about 2 kcal./mole and the change for $Fe(H_2O)_6^{+3}$ is zero.   There is no electron-pairing energy to be considered, and it is doubtful that the correlation energy will contribute much to this reaction.   Thus the contribution of LFSE, though small, will be significant in many reactions.   Indeed, in extreme cases the change of LFSE may be sufficient to change the distribution of the electrons among the d orbitals and the magnetic properties of the reacting ions.   More consideration should be given to these aspects in the future than they have received in the past, especially when the activation energy is small.

DISTORTED COMPLEX IONS

There is an additional potential source of stabilization energy in a complex ion of high symmetry.   A complex ion with a threefold, or higher, symmetry axis is potentially able to distort to a nuclear configuration of lower symmetry and lower energy.[50]   There will be more than one distortion possible that can lower the energy of the complex ion.   If the potential-energy barriers separating the nuclear configurations of the different distortions are low, the ion will undergo a pseudo-rotation from one nuclear configuration to another. If the potential-energy barriers are sufficiently high, the molecule will distort to one nuclear configuration and essentially remain there.   If the complex ion is nonlinear and electronically degenerate, the distortions are called the *dynamic* and *static Jahn-Teller effects*, respectively.   Liehr[50] has recently shown that electronically nondegenerate, nonlinear molecules may also show this behavior.   When the potential-energy barriers between nuclear configurations are high and the molecule is statically distorted, evidence on the nature of the distortion can be obtained from such sources as anisotropic g values, spectra, and crystal structures.   However, distortions observed in the solid state may be due to effects of charge, size, and polarizability on the crystal packing rather than to the symmetry-reducing forces referred to above. The extent of the distortion of a complex ion in the solution phase is unknown, but assumed small, for most ions with the possible exception of $Cr^{+2}(aq)$ and $Cu^{+2}(aq)$.   Where distortion occurs, a corrective term must be added to the ΔH and ΔF expressions for the high symmetry configuration of the complex ion.

π-BONDING IN COMPLEXES

The presence of π-bonding between the central metal ion and the ligands will affect the potential energy of the complex by increasing the bond energy of the metal-ligand bond and decreasing the metal-ligand bond length. The central metal ion may have vacant orbitals and receive π-bonding electrons from the ligand, or it may provide π-bonding electrons to empty ligand orbitals. Ligands such as OH^-, O^{-2}, and Cl^-, to mention only a few, can furnish p electrons to vacant metal d orbitals to make π bonds. Likewise, a central metal ion with electron pairs available in the d_{xy}, d_{xz}, and d_{yz} orbitals can furnish electrons to vacant ligand orbitals of the correct symmetry. In either case, delocalized electrons may be spread over the metal ion and at least a portion of the ligands. Overlap between two such π bonds, one on each complex ion, may provide an easy path for electron transfer between two complexes. This is especially true for OH^-, O^{-2}, and Cl^- bridges in inner-sphere activated complexes. It is also true for outer-sphere activated complexes such as the one which would be formed between the tris-1,10-phenanthroline Fe(II) and Fe(III) complexes in reaction (1–1). The K^+ ion in the bridged outer-sphere activated complex of reaction (1–6) may also facilitate electron transfer between the ferrocyanide and ferricyanide ions because the empty $3d$ or $4p$ orbitals on K^+ may provide weak π bonds with π molecular orbitals of the complex ions.

ENTROPY OF IONS IN WATER

This subject has been reviewed recently by Hunt.[35]   Here it will only be noted that two empirical equations have been proposed[51,52,53] for the absolute partial molal entropies of aqueous ions; these are:

$$\bar{S}^\circ_{abs} = \frac{3}{2} R \ln M + 10.2 - 11.6 \frac{Z^2}{r_u}$$

and

$$\bar{S}^\circ_{abs} = \frac{3}{2} R \ln M + 36.5 - 322 \frac{Z}{r_e^2}$$

where r_u is Pauling's univalent radius[54] and r_e is an effective radius.

These empirical expressions are useful when comparing the entropies of activated complexes, given by

$$S^{\ddagger} = \sum \bar{S}^\circ_{abs} + \Delta S^{\ddagger}$$

where the sum is taken over all reactants, with charges on the activated complexes.

3

Electron-Transfer Reactions in Homogeneous Solutions

This chapter will be devoted mainly to a discussion of reaction mechanisms which have been or may be postulated to explain experimental results obtained in studies of certain oxidation-reduction reactions involving the transfer, real or apparent, of electrons from one chemical species to another in solution. A relatively few examples will be discussed. Space does not permit a critical and complete discussion of all the many electron-transfer reactions which have been studied; an uncritical presentation of many studies is not particularly enlightening.

THE McKAY EQUATION

A number of the reactions to be discussed are isotope exchange reactions of the type[55]

$$AX + B \, *X \rightleftharpoons A \, *X + BX$$

in which there is only one group (ion, atom) X being exchanged between two chemicals AX and BX. An isotopic tracer, *X, which can be differentiated from X by some analytical procedure, is introduced into the system, either in AX, in BX or in both, to follow the rate of exchange. Neglecting differences in zero-point energies due to differences of isotope mass, *X is distributed statistically through the system according to the relative concentrations of total AX and BX forms. The net rate of appearance of *X in A *X is given by

$$\frac{dx}{dt} = R \frac{y}{b} \frac{(a - x)}{a} - R \frac{x}{a} \frac{(b - y)}{b} \tag{3-1}$$

in which $x = [A \, *X]$, $a = [AX] + [A \, *X]$, $y = [B \, *X]$, $b = [BX] + [B \, *X]$ and R is the rate at which X (all isotopes) is exchanged between AX and BX. The applicability of Eq. 3–1 does not depend on the form of R, but it does assume that isotopic effects do not effect R, e.g., it assumes that the specific rate constant(s) involved in R is the same for the exchange of *X between the two species as for the exchange of X between the two species. The ratio y/b gives the fraction of the exchanges which involve B *X and $y(a - x)/ab$ gives the fraction of the exchanges which simultaneously involve B *X and AX so that transfer of *X from B *X to AX is accomplished. Similarly, $x(b - y)/ab$ gives the fraction of the exchanges which simultaneously involve A *X and BX so that the reverse reaction transferring *X from A *X to BX is accomplished. Thus Eq. 3–1 gives the net rate of transfer of *X from the BX species to the AX species.

In most cases the quantity of tracer is much less than the total amount of the substance present, so that $(a - x) \cong a$, $(b - y) \cong b$, and Eq. 3–1 becomes

$$\frac{dx}{dt} = R\left(\frac{y}{b} - \frac{x}{a}\right) \tag{3–2}$$

If the tracer *X is not so unstable as to appreciably change its quantity during the time of an experiment, then

$$x + y = \text{a constant} \tag{3–3}$$

With this relationship, and assuming a and b to be constant, Eq. 3–2 may be integrated to give

$$\ln \frac{x - x_\infty}{x_0 - x_\infty} = -R\frac{a + b}{ab}t \tag{3–4}$$

where x_0, x, and x_∞ are the concentrations of A *X (the radioactivity counts of AX may be used in place of concentrations) at zero time, time t, and infinite time, respectively. For $x_0 = 0$, Eq. 3–4 reduces to

$$\ln(1 - F) = -R\left(\frac{a + b}{ab}\right)t \tag{3–5}$$

which is usually called the McKay equation. Here $F = x/x_\infty$, and is the fractional exchange at time t.

A small quantity of AX (containing A *X but free from BX) is isolated at time t and at "infinite time," usually eight to ten half-times of the exchange. A determination of the relative amounts of A *X in the two samples yields a value of F for time t. Samples taken at various times t during the exchange permit $\ln(1 - F)$ to be plotted versus t. The slope of the line so obtained is equal to $-R(a + b)/ab$ and is used to obtain a value of R for the reaction conditions employed. By varying the reaction conditions, the functional dependence of R upon concentration, temperature, and other experimental variables may be determined.

If exchange is somehow induced during the isolation of AX + A *X, or if BX (containing B *X) contaminates the sample of AX, then an equation similar to Eq. 3–5 can be used when these sources of error are reproducible; it is[56]

$$\ln(1 - F') = \ln(1 - F_0) - R\left(\frac{a + b}{ab}\right)t \tag{3–6}$$

in which F' and F_0 are observed fractions of exchange after separation at time t and zero time, respectively. As before, the slope of a plot of $\ln(1 - F')$ versus t yields a value of R for a particular set of experimental conditions.

THE Fe(II) + Fe(III) ISOTOPE EXCHANGE REACTION

The first studies of this exchange gave very contradictory results,[57,58,59] which were resolved by the study of Silverman and Dodson.[60] These investigators employed Fe(II) and Fe(III) concentrations of the order of $10^{-4}M$, and separated the Fe(II) and Fe(III) species by adding aliquots of reaction mixture to a solution of 2,2'-bipyridine in an acetic acid–acetate buffer of pH 5 and precipitating hydrous ferric oxide with ammonia. The Fe(II) remained in solution as the $Fe(bipy)_3^{+2}$ complex. The activities of ^{55}Fe in the ferric oxide precipitates were determined. Zero-time exchange was approximately 35 percent. From variations of [Fe(II)], [Fe(III)], and [H^+], the rate of exchange was found to be given by

$$R = \left\{ k_0 + \frac{k'}{[H^+]} \right\} [Fe(II)][Fe(III)] \tag{3-7}$$

Hence exchange by the following mechanism:

$$*Fe^{+2}(aq) + Fe^{+3}(aq) \xrightarrow{k_0} *Fe^{+3}(aq) + Fe^{+2}(aq) \tag{3-8}$$

$$*Fe^{+2}(aq) + FeOH^{+2}(aq) \xrightarrow{k_h} *FeOH^{+2}(aq) + Fe^{+2}(aq) \tag{3-9}$$

was postulated. According to this mechanism the rate of isotope exchange, R, is given by

$$R = k_0[Fe^{+2}][Fe^{+3}] + k_h[Fe^{+2}][FeOH^{+2}] \tag{3-10}$$

When the concentrations of Fe^{+3} and $FeOH^{+2}$ are expressed in terms of the total Fe(III) concentration, using Eqs. 3–11 and 3–12,

$$K_{1h} = \frac{[FeOH^{+2}][H^+]}{[Fe^{+3}]} \tag{3-11}$$

$$Fe[(III)] = [Fe^{+3}] + [FeOH^{+2}] \tag{3-12}$$

relation 3–13 is obtained for R:

$$R = \left\{ \frac{k_0[H^+] + k_h K_{1h}}{[H^+] + K_{1h}} \right\} [Fe^{+2}][Fe(III)] \tag{3-13}$$

Since the value of K_{1h} was taken to be 0.26×10^{-3} mole/l. at 0°C and $\mu = 0.55$, and since pH was in the range 0.26 to 1.79, it is seen that $[H^+] \gg K_{1h}$, so that Eq. 3–13 reduces to Eq. 3–7 with $k' = k_h K_{1h}$.

Exchange paths involving $Fe(OH)_2^+$ or the dimer $Fe_2(OH)_2^{+4}$ contributed negligibly since no terms for them are found in Eq. 3–7.

It is usually assumed that the exchange path involving OH^- is symmetrical.

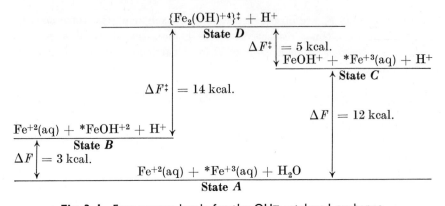

Fig. 3–1. Free-energy levels for the OH⁻ catalyzed exchange.

Electron transfer occurs either in a hydroxo-bridged inner-sphere activated complex

$$\{[\text{Fe} \cdots \text{OH}^- \cdots {}^*\text{Fe}]^{+4}\}^{\ddagger}$$

or by hydrogen atom transfer in a hydrogen-bridged outer-sphere activated complex

$$\left\{ \left[\begin{array}{ccc} \text{Fe—O} \cdots \text{H} \cdots \text{O—} {}^*\text{Fe} \\ \mid \quad\quad\quad\quad \mid \\ \text{H} \quad\quad\quad \text{H} \end{array} \right]^{+4} \right\}^{\ddagger}$$

In either case the products and reactants are identical, as shown in reaction (3–9), and k_h can be calculated from $k_h = k'/K_{1h}$. However, it should not be forgotten that the possibility of electron transfer by the unsymmetrical path

$$\text{Fe}^{+2}(\text{aq}) + {}^*\text{FeOH}^{+2} \rightleftharpoons \text{X}^{\ddagger} \rightleftharpoons \text{Fe}^{+3}(\text{aq}) + {}^*\text{FeOH}^+ \qquad (3\text{–}14)$$

still exists. If this path is involved, the electron transfer would presumably occur in an outer-sphere activated complex not involving simultaneous transfer of a proton. Because the concentrations of all the chemical species remain constant while *Fe is being distributed, the principle of detailed balance states that the rates of reaction between $\text{Fe}^{+2}(\text{aq})$ and FeOH^{+2} and between FeOH^+ and $\text{Fe}^{+3}(\text{aq})$ are equal. Consider Fig. 3–1. From the free-energy differences between states A, B, C, and D, it is seen that the rates of reaction for these two pairs of reactants are given by

$$R_1 = k_h[\text{Fe}^{+2}][\text{FeOH}^{+2}] = \frac{kT}{h} \exp\left(\frac{-14{,}000}{RT}\right) \exp\left(\frac{-3000}{RT}\right) \frac{[\text{Fe}^{+2}][\text{Fe}^{+3}]}{[\text{H}^+]}$$
$$(3\text{–}15)$$

$$R_2 = k_h'[\text{FeOH}^+][\text{Fe}^{+3}] = \frac{kT}{h} \exp\left(\frac{-5000}{RT}\right) \exp\left(\frac{-12{,}000}{RT}\right) \frac{[\text{Fe}^{+2}][\text{Fe}^{+3}]}{[\text{H}^+]}$$
$$(3\text{–}16)$$

and that $R_1 = R_2$.

Since both K_{1h} and K'_{1h}, the hydrolysis constant for Fe^{+2}, are very much less than $[H^+]$, $[Fe^{+3}] \cong [Fe(III)]$, $[Fe^{+2}] \cong [Fe(II)]$, and R is given by

$$R = \left\{ k_0 + \frac{(k_h K_{1h} + k'_h K_{1h})}{[H^+]} \right\} [Fe(II)][Fe(III)] \qquad (3\text{--}17)$$

The observed rate constant is given by

$$k = k_0 + \frac{2 k_h K_{1h}}{[H^+]} \qquad (3\text{--}18)$$

In this case k' in Eq. 3–7 is equal to $2 k_h K_{1h}$ and k_h is equal to $k'/2K_{1h}$, a value only half as large as the one now attributed to k_h.

The enthalpies and entropies of activation of reactions (3–8) and (3–9) reported by Fukushima and Reynolds[61] are $\Delta H_0^{\ddagger} = 10.5 \pm 0.85$, $\Delta H_H^{\ddagger} = 8.44 \pm 0.47$ kcal./mole and $\Delta S_0^{\ddagger} = -20.6 \pm 3.0$, $\Delta S_H^{\ddagger} = -14.4 \pm 1.6$ cal./deg. mole, respectively, and are equal to the values given by Silverman and Dodson[60] within the experimental errors of the latter. The two enthalpies of activation have values which lie within the range of $\Delta H^{\ddagger}$ values for exchange paths involving ligands which form labile complexes with Fe^{+2} and Fe^{+3}. (See Table 3–1.)

Effect of Other Ligands on Exchange Rate

When Fe^{+3} can form, in a stepwise manner, the complex ions FeX_n^{+3-nz} ($n = 1, 2, 3, \ldots$) where z is the absolute value of the valence of the ligand, the rate of exchange can be written as

$$R = k[Fe(II)][Fe(III)]$$

$$= \left\{ \frac{k_0 + (k_h K_{1h}/[H^+]) + k_1 K_1 [X^{-z}] + k_2 K_1 K_2 [X^{-z}]^2 + \cdots}{1 + (K_{1h}/[H^+]) + K_1 [X^{-z}] + K_1 K_2 [X^{-z}]^2 + \cdots} \right\} [Fe(II)][Fe(III)]$$
$$(3\text{--}19)$$

where the k_i are the rate constants for reaction between $Fe^{+2}(aq)$ and FeX_i^{+3-iz}. In deriving this equation, it was assumed that the pH was such that no more than the first hydrolysis product of Fe^{+3} was formed, that hydrolysis and complexing of Fe^{+2} was unimportant, and that the complexes of Fe(III) do not explicitly involve H^+ ions. These assumptions are usually valid. It was also assumed that the FeX_n^{+3-nz} complexes were rapidly formed in the equilibrium

$$Fe^{+3} + nX^{-z} = FeX_n^{+3-nz} \qquad (3\text{--}20)$$

so the rate-determining step was the reaction between $Fe^{+2}(aq)$ and FeX_n^{+3-nz}. Measurements of forward and reverse rate constants for reaction (3–20) for $n = 1$, $X^{-z} = SCN^-$, Cl^-, F^- have been made.[62,63,64] For both SCN^- and Cl^-, the rate of complex formation is given by

$$R_c = \{ k_1 + (k_2/[H^+]) \} [Fe^{+3}][X^-] \qquad (3\text{--}21)$$

TABLE 3-1

Fe(II) + Fe(III) Electron Exchange Reactions

Reaction	μ	$T°C$	k $\left(\dfrac{\text{liter}}{\text{mole sec.}}\right)$	$\Delta H\ddagger$ $\left(\dfrac{\text{kcal.}}{\text{mole}}\right)$	$\Delta S\ddagger$ $\left(\dfrac{\text{cal.}}{\text{deg. mole}}\right)$	Ref.
$Fe^{+2} + Fe^{+3}$	0.55	0	0.87	9.3	-25	60
$Fe^{+2} + Fe^{+3}$	0.50	25	4.0 ± 0.4	10.5 ± 0.85	-20.6 ± 3.0	61
$Fe^{+2} + FeOH^{+2}$	0.55	0	1×10^3	6.9	-18	60
$Fe^{+2} + FeOH^{+2}$	0.50	25	$(2.99 \pm 0.09)10^3$	8.44 ± 0.47	-14.4 ± 1.6	61
$Fe^{+2} + Fe^{+3}(D_2O)$	0.50	7	0.7	$\pm$	—	78
$Fe^{+2} + Fe^{+3}(D_2O)$	0.50	5	0.40 ± 0.01	—	—	61
$Fe^{+2} + FeOH^{+2}(D_2O)$	0.50	7	765	—	—	78
$Fe^{+2} + FeOH^{+2}(D_2O)$	0.50	25	$(3.04 \pm 0.16)10^3$	11.5 ± 0.5	-4.0 ± 1.8	61
$Fe^{+2} + FeF^{+2}$	0.50	0	9.7	8.6	-21	66
$Fe^{+2} + FeF_2^+$	0.50	0	2.5	9.0	-22	66
$Fe^{+2} + FeF_3$	0.50	0	0.5	—	—	66
$Fe^{+2} + FeCl^{+2}$	0.55	20	29	8.3	-24	60
$Fe^{+2} + FeCl^{+2}$	0.50	20	22.8	11.0	-15	79
$Fe^{+2} + FeCl^{+2}(D_2O)$	0.50	20	9.1	13.2	-10	79
$Fe^{+2} + FeCl^{+2}$	3.0	25	57.6 ± 2	—	—	69
$Fe^{+2} + FeCl_2^+$	0.55	20	53	9.5	-20	60
$Fe^{+2} + FeCl_2^+$	3.0	25	159 ± 10	—	—	69
$Fe^{+2} + FeBr^{+2}$	0.50	0	4.9	8.0	-25	70
$Fe^{+2} + FeBr_2^+$		0	19	14	-3	70
$Fe^{+2} + FeSCN^{+2}$		0	4.2	9.2	-21	70
$Fe^{+2} + FeSCN^{+2}$	0.50	0	12.2	7.4	-26	65
$Fe^{+2} + Fe(SCN)_2^+$		0	41.2	10.8	-13	70

Reaction	μ	t, °C	k	E	ΔS‡	Ref.
$Fe^{+2} + Fe(SCN)_2^+$	0.50	0	2.0	8.1	−28	65
$Fe^{+2} + FeN^{+2}$	0.55	10	4.75×10^3	13.3	+7	25
$Fe^{+2} + FeN_3^{+2}(D_2O)$	0.55	10	3.81×10^3	14.0	+8	25
$Fe^{+2} + FeN_3^{+2}(H_2O)$	0.55	27	10.3×10^3	~5.5	−24	25
$Fe^{+2} + FeC_2O_4^+$	0.55	0	7.0×10^2	8.7	−14	29
$Fe^{+2} + FeC_2O_4^+$		0	2.5×10^2	21	+28	71
$Fe^{+2} + Fe(C_2O_4)_2^-$	0.55	0	3.6×10^3	—	—	29
$Fe^{+2} + FeSO_4^+$	0.25	25	692	—	—	72
$Fe^{+2} + FeSO_4$	0.50	25	346	8.4	−19	73
$Fe^{+2} + FeSO_4^+$	1.0	25	295	13.2	−3	74
$Fe^{+2} + FeSO_4^+$	1.0	28	546	13	−2	71
$Fe^{+2} + Fe(SO_4)_2^-$	0.25	25	1.94×10^4	—	—	72
$Fe^{+2} + Fe(SO_4)_2^-$	1.0	25	1.75×10^4	15	+9	74
$FeSO_4 + FeOH^{+2}$	0.25		2×10^6	—	—	72
$Fe^{+2} + FeHPO_4^+$		0	5.2×10^2	15	+6	71
$Fe^{+2} + Fe(EDTA)^-$		~25	$<4 \times 10^{-4}$	—	—	86
$Fe^{+2} + Fe(bipy)_3^{+3}$		25	2.7×10^4	—	—	87
$Fe^{+2} + Fe(tripy)^{+3}$		25	8.5×10^4	—	—	87
$Fe^{+2} + Fe(ph)_3^{+3}$		25	3.7×10^4	0.2	−37	88
$Fe^{+2} + Fe(5\text{-}Me\text{-}ph)_3^{+3}$		25	2.0×10^4	—	—	87
$Fe^{+2} + Fe(5\text{-}NO_2\text{-}ph)_3^{+3}$		25	1.1×10^6	—	—	87
$Fe^{+2} + Fe(5\text{-}Cl\text{-}ph)_3^{+3}$		25	2.1×10^5	—	—	87
$Fe(ph)_3^{+2} + Fe(ph)_3$		0	$>10^5$	—	—	89
$Fe(d\text{-}ph)_3^{+2} + Fe(d\text{-}ph)_3^{+3}$		0	$>2 \times 10^3$	—	—	18
$Fe(d\text{-}dipy)_3^{+2} + Fe(ph)_3^{+3}$		25	$>10^8$	—	—	90
$Fe(t\text{-}ph)_3^{+2} + Fe(d\text{-}dipy)_3^{+3}$		25	$>10^8$	—	—	90
$Fe(CN)_6^{-4} + Fe(ph)_3^{+3}$		25	$>10^8$	—	—	90
$Fe(CN)_6^{-4} + Fe(CN)^{-3}(0.01\ M\ KOH)$	0.1	26	355	4.1	−32	91
$Fe(C_5H_5)_2 + Fe(C_5H_5)_2^+$			10^5	—	—	76

At 0°C and 0.0967 M H^+, where a number of Fe^{+2} + $*Fe^{+3}$ + SCN^- isotope exchange measurements were made,[65] $R_c = 23.9[Fe^{+3}][SCN^-]$ mole/l. sec. At 0°C and 0.547 M H^+ for the Cl^--catalyzed isotope exchange path,[60] $R_c = 1.48[Fe^{+3}][Cl^-]$ mole/l. sec. The rates of the isotope exchange reactions involving $(Fe_2SCN)^{+4}$ and $(Fe_2Cl)^{+4}$ at the same temperature and pH are

$$R_{ex}(SCN^-) = 4.15 \times 10^3[Fe^{+2}][Fe^{+3}][SCN^-] \text{ mole/l. sec.}$$

$$R_{ex}(Cl^-) = 20[Fe^{+2}][Fe^{+3}][Cl^-] \text{ mole/l. sec.}$$

respectively. Since $[Fe^{+2}] \simeq (2 \text{ to } 5) \cdot 10^{-4} M$ in most exchange experiments, $R_c > R_{ex}$ in both cases.

However,[64] in the case of F^- at 0° and 0.4 M H^+, $R_c = 762[Fe^{+3}][F^-]$ mole/l. sec., whereas $R_{ex} = 9.4 \times 10^5[Fe^{+2}][Fe^{+3}][F^-]$ mole/l. sec. Hudis and Wahl[66] used iron concentrations where $R_{ex} < R_c$. When the Fe^{+2} concentration is increased to $10^{-3} M$, or greater, $R_c < R_{ex}$ and the rate of exchange between Fe(II) and Fe(III) is consistent with the following mechanism:

$$Fe^{+2}(aq) + *Fe^{+3}(aq) \xrightarrow{k_0} Fe^{+3}(aq) + *Fe^{+2}(aq) \qquad (3\text{–}8)$$

$$Fe^{+2}(aq) + *FeOH^{+2} \xrightarrow{k_h} FeOH^{+2} + *Fe^{+2}(aq) \qquad (3\text{–}9)$$

$$Fe^{+2}(aq) + *FeF^{+2} \xrightarrow{k_1} FeF^{+2} + *Fe^{+2}(aq) \qquad (3\text{–}22)$$

$$F^- + *Fe^{+3}(aq) \xrightarrow{k_{f1}} *FeF^{+2} \qquad (3\text{–}23)$$

$$HF + *Fe^{+3}(aq) \xrightarrow{k_{f2}} *FeF^{+2} + H^+ \qquad (3\text{–}24)$$

The fraction exchange at time t is given by

$$1 - F = \Theta e^{-\alpha t} + (1 - \Theta)e^{-\beta t} \qquad (3\text{–}25)$$

in which

$$\Theta = \frac{1}{(\beta - \alpha)}\left\{\frac{b}{b + c}\left(\frac{R_2}{c} - \frac{R_1}{b}\right) + \left(\beta - \frac{R_1 + R_2}{a} - \frac{R_2}{c}\right)\right\}$$

$$\alpha = \tfrac{1}{2}(p - \sqrt{p^2 - 4q})$$

$$\beta = \tfrac{1}{2}(p + \sqrt{p^2 - 4q})$$

$$p = \frac{a + b}{ab}R_1 + \frac{a + c}{ac}R_2 + \frac{b + c}{bc}R_3$$

$$q = \frac{(a + b + c)}{abc}(R_1 R_2 + R_1 R_3 + R_2 R_3)$$

$$R_1 = (k_0 + k_h K_{1h}/[H^+])[Fe^{+2}][Fe^{+3}]$$

$$R_2 = k_1[Fe^{+2}][FeF^{+2}]$$

$$R_3 = (k_{f1} + k_{f2}[H^+]/K_{HF})[Fe^{+3}][F^-]$$

$$a = [Fe^{+2}] + [*Fe^{+2}] \simeq [Fe^{+2}]$$

$$b = [Fe^{+3}] + [*Fe^{+3}] \simeq [Fe^{+3}]$$

$$c = [FeF^{+2}] + [*FeF^{+2}] \simeq [FeF^{+2}]$$

If isotope exchange occurs between Fe^{+2}(aq) and $Fe(H_2O)_6F^{+2}$, an outer-sphere complex ion of F^- and $Fe(H_2O)_6^{+3}$, according to

$$Fe^{+2}(aq) + {}^*Fe(H_2O)_6F^{+2} \xrightarrow{k_{26}} Fe(H_2O)_6F^{+2} + {}^*Fe^{+2}(aq) \qquad (3\text{-}26)$$

or through a termolecular reaction with an outer-sphere activated complex, then the expression for R_1 would have the additional term $k'_{26} K'_1 [F^-][Fe^{+2}]$ $[Fe^{+3}]$ in which k'_{26} is a rate constant for, and K'_1 is a formation constant of, the outer-sphere complex $Fe(H_2O)_6F^{+2}$. It has been found[67] that the observed rate of exchange is in very good agreement with that calculated from Eq. 3–25 when the appropriate data for the various rate constants and equilibrium constants at 0.400 M H^+ and 0°C were used, omitting reaction (3–26). Thus there is good evidence that, under the conditions at which F^- catalysis of the Fe(II) + Fe(III) exchange was studied, only the inner coordination sphere complex FeF^{+2} contributed appreciably to the exchange rate. The actual electron-transfer mechanism could either be through an inner-sphere activated complex with F^- or H_2O bridge or through an unsymmetrical outer-sphere activated complex. The inner-sphere activated complex with F^- would seem to be more probable than the inner-sphere activated complex with H_2O bridge, but the data did not permit a choice between the inner-sphere and outer-sphere activated complexes. On referring to Fig. 3–1, it is seen that $\Delta F = 12$ kcal. for replacing a water molecule with OH^- on Fe^{+2}(aq); the over-all ΔF accompanying the replacement of H_2O with FeF^{+2} in formation of the bridged inner-sphere activated complex would be greater than this, in agreement with observed results.

In the oxidation[68] of Cr^{+2}(aq) to Cr(III) by Fe^{+3} in the presence of Cl^- electron transfer apparently occurs in both the Cl^--bridged inner-sphere activated complex and activated complexes not involving the inner-sphere $FeCl^{+2}$ complex ion. When Cl^- was present in the Fe^{+3} stock solution and the initial concentration of $FeCl^{+2}$ was larger than that of Cr^{+2}(aq), 99 percent of the Cr(II) was converted to inert $CrCl^{+2}$, indicating that Cl^- was in the inner coordination sphere of Cr(II) when it was oxidized to Cr(III). When Cl^- was not initially present in the Fe^{+3} stock solution, two electron transfer paths, with rates given by $k[Fe^{+3}][Cl^-]$ and $k'[Cl^-] \times [Fe^{+3}][Cr^{+2}]$, were found. The former corresponds to formation of $FeCl^{+2}$, which was reduced as rapidly as it was formed. The latter corresponds to a reaction with Cl^- in an outer coordination sphere of $Fe(H_2O)_6^{+3}$. The product Cr(III) contained 31 percent $CrCl^{+2}$ with approximately 12 percent coming from formation of $FeCl^{+2}$ and the rest coming from a path involving an outer-sphere activated complex $[(H_2O)_5Fe(H_2O)ClCr(H_2O)_5]^{+4}$ or a water-bridged inner-sphere activated complex $[(H_2O)_5Fe\text{-}(H_2O)\text{-}Cr(HO_2)_4Cl]^{+4}$, or both.[68] The product Cr(III) also contained 69 percent $Cr(H_2O)_6^{+3}$ which came from the reaction of Cr^{+2}(aq) with Fe^{+3}(aq) and with $FeOH^{+2}$.

By comparing the rate constant for the Fe(II)-catalyzed dissociation of

$FeCl^{+2}$ with the rate constant for the $Fe^{+2} + FeCl^{+2}$ exchange path, Campion, Conocchioli, and Sutin[69] found that the latter reaction occurred by a Cl^--bridged inner-sphere mechanism and by either a H_2O-bridged inner-sphere mechanism or an outer-sphere mechanism.

When $n = 1$, $X^{-z} = N_3^-$, it seems that the transition state is not formed directly by reaction of $Fe^{+2}(aq)$ and FeN_3^{+2} (see Ref. 25). A dimer is first formed with the empirical formula $(Fe_2N_3)^{+4}$ and then the transition state for electron transfer may be formed from the dimer. At temperatures below 15°C, the formation of dimer is rate-determining; at temperatures above 15°C, the rearrangement of ligands in the dimer is rate-determining. Both reactions have the same concentration dependence $[Fe^{+2}][Fe^{+3}][N_3^-]$ but different T dependence. At low T, dimer formation, with a large $\Delta H^{\ddagger}$, is rate-determining; at higher T, rearrangement, with a smaller $\Delta H,^{\ddagger}$ is rate-determining. Thus the existence of a stable bridged dimer with bonds of considerable strength is not sufficient, by itself, to ensure rapid electron transfer. This case emphasizes the importance of the energy-balance requirement in bridged inner-sphere activated complexes.

Many other anions also catalyze the $Fe(II) + Fe(III)$ isotope exchange. Exchange paths involving two Cl^- (see Ref. 60), two and three F^- (see Ref. 66), Br^- (see Ref. 70), $C_2O_4^{-2}$ (see Refs. 29, 71), SCN^- (see Refs. 65, 70), SO_4^{-2} (see Refs. 70, 71, 72, 73, 74), HPO_4^{-2} (see Ref. 71), tartrate ion,[75] or $C_5H_5^-$ (see Ref. 76) are known. Rate constants and enthalpies and entropies of activation, when known, are listed in Table 3–1. Other reactions involving ethylenediaminetetracetate (EDTA), CN^-, and 1,10-phenanthroline (ph) complexes are also listed in Table 3–1 but are treated in a later section.

Rate constants for the various exchange reactions have not been obtained for infinite dilution; for greater rigor in making comparisons, they should be. However, more serious sources of error exist because results reported by different investigators often show considerable discrepancies. For example, Horne[29] reported $E_{act} = 9.2$ kcal./mole whereas Sheppard and Brown[71] reported $E_{act} = 21$ kcal./mole for the single $C_2O_4^{-2}$-catalyzed path. Willix[73] reported 8.3, Sheppard and Brown[71] reported 13.5, and Bachmann and Lieser[74] reported 13.8 kcal./mole for E_{act} for the single SO_4^{-2} path at $\mu = 1$. The rate constants k_1 reported for this path are 346,[73] 295,[74] 515 (calculated from the data of Sheppard and Brown, using improved values of K_1 (see Ref. 73) and K_a for HSO_4^- (see Ref. 72) in $NaClO_4$–$HClO_4$–H_2SO_4 solutions), and approximately 540 l./mole sec. (extrapolated value for $\mu = 1$ from the data of Reynolds and Fukushima[72] at various μ). The value of 346 l./mole sec. given by Willix is suspect since he neglects the contribution to the exchange rate from the disulfato path. At a total acid concentration of 0.400 M and a total sulfate concentration of 0.0558 M, it is readily calculated that free SO_4^{-2} concentration is 0.006 M, using $K_a = 0.042$ M for HSO_4^- given by Willix, which is in good agreement with the value which may be calculated from the log $K_a = f(\mu)$ expression of Reynolds and Fukushima[72] but in poor agreement with the value of 0.37 used by Sheppard and Brown.[71] Thus the rate of the monosulfato path is

$$R_1 = k_1 K_1 [SO_4^{-2}][Fe^{+2}][Fe^{+3}] = 415[Fe^{+2}][Fe^{+3}] \ M/sec.$$

(using Willix' value $K_1 = 200 \ M^{-1}$, which is in agreement with the equilibrium constants of Mattoo[77] but in serious disagreement with the value of $25 \ M^{-1}$ used by Sheppard and Brown). If one accepts the value of approximately 2×10^4 l./mole sec. found by Reynolds and Fukushima[72] and by Bachmann and Lieser[74] for k_2, and of $5 \ M^{-1}$ for K_2 from Mattoo,[77] it is readily calculated that the rate of exchange by the disulfato path is approximately

$$R_2 = k_2 K_1 K_2 [SO_4^{-2}]^2 [Fe^{+2}][Fe^{+3}] = 720[Fe^{+2}][Fe^{+3}] \ M/sec.$$

Hence isotope exchange by the disulfato path under the conditions used by Willix is at least as important as exchange by the monosulfato path.

Reynolds and Fukushima[72] found an exchange path in the SO_4^{-2} catalysis not represented in Eq. 3–19, namely, one involving both one SO_4^{-2} and one OH^- so that the coefficient of $[SO_4^{-2}]$ in the rate expression was linearly dependent on $[H^+]^{-1}$. The decrease of free energy caused by adding one OH^- to the monosulfato activated complex is very nearly equal to the decrease of free energy caused by adding one OH^- to the aquo activated complex $[Fe_2^{+5}]^{\ddagger}$. This suggests that OH^- plays the same role in the activated complexes $[(Fe_2(SO_4)OH^{+2}]^{\ddagger}$ and $[Fe_2OH^{+4}]^{\ddagger}$. Perhaps activated complexes analogous to the hydroxomonosulfato one would be found in some of the other anion catalyses if the rate measurements were performed with greater precision.

Effect of D₂O

The effect of D_2O on Fe(II) + Fe(III) exchanges has been studied.[78,79,61] In the first two studies it was found that the rate-constant values in D_2O were approximately one-half the rate-constant values in H_2O. A study of the D_2O effect as a function of temperature[61] in perchlorate media showed that $\Delta H^{\ddagger}$ was 3.1 ± 0.7 kcal./mole greater and that $\Delta S^{\ddagger}$ was 10.0 ± 2.4 cal./deg. mole greater in D_2O than in H_2O for the OH^--catalyzed path. At 25°C the rate constant k_h was virtually the same in both H_2O and D_2O. This result points out the danger of making measurements of the D_2O effect at only one temperature—if the D_2O effect had been measured at 25°C instead of 7°C it might have been concluded that there was no D_2O effect on k_h. At temperatures above 25°C, presumably the rate constant is greater in D_2O than in H_2O and the isotope effect is reversed if the mechanism of the reaction does not change.

The D_2O effect was first thought to be evidence for the making and breaking of O—H bonds in the rate-determining step, for example, in hydrogen-atom transfer. It is now clearly recognized that D_2O effects can arise from bridge water molecules in inner-sphere activated complexes as in the $Co(NH_3)_5OH_2^{+3} + Cr^{+2}(aq)$ reaction or from solvent rearrangement[80] without the actual rupture of hydrogen-containing bonds as in the $Cr(NH_3)_5Cl^{+2} + Cr^{+2}(aq)$ reaction[81] which proceeds via a Cl^--bridged inner-sphere activated complex. Even though no bond containing a hydrogen atom is ruptured, D_2O effects in reactions must be expected because ions, with different enthalpies and entropies of hydration in H_2O and D_2O, appear and disappear in a reaction. The equilibrium constant K_1 for $Fe^{+3} + Cl^- = FeCl^{+2}$ is doubled in D_2O as compared to H_2O[79] although no net transfer of hydrogen atoms between ions occurs.

Effect of Ice-like Media

An interesting set of experiments with considerable potential has been conducted by Horne[82] in his investigation of the rate of the Fe(II) + Fe(III) isotope exchange reaction at very low temperatures in H_2O–$HClO_4$ liquid and solid media. In solid ice-like media the electron was apparently transferred over distances of the order of 100 Å. The rate was approximately first-order in Fe(II) and Fe(III) concentrations and was not diffusion-controlled. The activation energy was 8.4 kcal./mole, approximately the same value as for liquid H_2O–$HClO_4$ media. Grotthuss-type conduction of hydrogen atoms along a chain of oriented water molecules has been suggested as a possible mechanism for electron transfer.[34]

Great care must be exercised in planning rate experiments in solid media. Separation of phases with consequent changes of concentrations will be a major concern, at least until much more is known concerning the magnitude of this possible source of error. Use of eutectic mixtures which solidify with the same composition as the liquid phase will help but will impose great limitations on the variation of the supporting electrolyte concentration. Reaction mixtures quickly solidified when applied as a very fine spray to a very cold surface may give reproducible solid reaction media with a chance to vary the supporting electrolyte.

Effect of Solvent

Studies of the Fe(II) + Fe(III) isotope exchange rate in non-aqueous media have been aimed at determining whether water in the inner spheres of the reactants is necessary to the electron-transfer process. Horne[70] used alcohol–water mixtures. For methanol, ethanol, and 1-propanol, the rate constant decreased with decreasing water content. For example, the ob-. served rate constant at 20.5°C, 0.548 M $HClO_4$, 1.06 × 10^{-4} M Fe(II) and

2.12×10^{-4} M Fe(III) was 8.22 M^{-1} sec.$^{-1}$ in absence of alcohol, 1.59 M^{-1} sec.$^{-1}$ in 0.58 MeOH, 1.21 M^{-1} sec.$^{-1}$ in 0.49 EtOH, and 1.8 M^{-1} sec.$^{-1}$ in 0.43 PrOH (the alcohol concentrations are mole fractions). At smaller water concentrations in ethyl alcohol the rate constants appeared to decrease continuously with decrease of water content. At least some water appeared to be necessary for electron transfer.

Sutin[83] found the observed rate in isopropyl alcohol, containing less than 0.01 M HClO$_4$, was 10^{-8} times smaller than the observed rate in 0.01 M aqueous HClO$_4$. Three major reasons were given[83] for the large change in rate. These were:

1. The nuclear configurations of the inner spheres of the reactants differ more in alcohol than in water requiring a larger reorganization energy.
2. Water may play a special role, such as net hydrogen-atom transfer, which is poorly played by alcohols.
3. The dielectric constant of the bulk phase was changed. If it is assumed that $Fe(H_2O)_6^{+2}$ and $Fe(H_2O)_6^{+3}$ (or $Fe(H_2O)_5OH^{+2}$) approach to 6.8 Å in the activated complex, the difference of the coulombic repulsion energies between reactants in the activated complexes in the two solvents (assuming no other sources) is

$$\ln\left(\frac{k_{alc}}{k_{H_2O}}\right) = \frac{z_1 z_2 e^2}{rkT}\left(\frac{1}{\varepsilon_{H_2O}} - \frac{1}{\varepsilon_{alc}}\right)$$

which yields approximately 10^{-10} to 10^{-11} for the ratio of the rate constants (or 10^{-7} if the charge product was 4 instead of 6).

Maddock[84] reported that in rigorously purified nitromethane the half-time for the exchange was many days at room temperature.

Investigations[85] in dimethylsulfoxide (DMSO) showed that, at total iron concentrations approximately equal to 10^{-4} M in NaClO$_4$–HClO$_4$ media, the exchange rate was first-order in Fe(II) and Fe(III) concentrations, that the rate in the absence of added water was independent of HClO$_4$ concentration in the range from zero to 0.15 M added HClO$_4$, and that the rate was independent of added water concentration up to at least 0.5 M. These results provided evidence that FeOH^{+2} and FeS^{+2}, where S$^-$ is a solvent anion, contributed negligibly in the absence of added water and that hydrated or partially hydrated $(Fe(DMSO)_{6-n}(H_2O)_n^{+3,+2})$ species had either negligible concentrations or had the same reaction rates as the solvated species $Fe(DMSO)_6^{+3,+2}$ within the limits studied. The rate constant observed at 20°C was $16 \pm 15\%$ l./mole sec. at $\mu = 0.2$.

Mechanisms of Electron Transfer

A few definite statements can be made about the mechanism of electron transfer in the uncatalyzed and anion-catalyzed ferrous-ferric exchange reactions. The principal mechanisms to be considered are:

1. Adiabatic electron-transfer in an outer-sphere activated complex which may, or may not, involve hydrogen bonds between the intact inner coordination spheres.
2. Electron transfer in a bridged inner-sphere activated complex.
3. Non-adiabatic electron transfer in an outer-sphere activated complex.
4. Net hydrogen-atom transfer along an ordered array of water molecules.

There is very little evidence that non-adiabatic electron transfer, with a transmission coefficient very much less than unity, is a rate-determining step in either an outer-sphere or a bridged inner-sphere activated complex. Sacher and Laidler[16] have shown that a tunnelling theory (see Chapters 1 and 5) can give an apparent $\Delta F^{\ddagger}$ value consistent with the experimental value. The iron-iron distance for a minimum $\Delta F^{\ddagger}$ was 4.2 Å; it is difficult to reconcile this short distance, which corresponds to a one-water-molecule bridge between the iron ions, with electron tunnelling as a slow step. At this small interionic distance the electronic interactions in a bridged dinuclear complex, with properly reorganized inner coordination and outer solvation shells, should permit rapid electron transfer. Values of $\Delta S^{\ddagger}$ in the range -15 to -25 cal./deg. mole are approximately the values expected for loss of motional entropy when an activated complex (either inner-sphere or outer-sphere) is formed in a bimolecular reaction and do not indicate a value of much less than unity for the transmission coefficient.

It was suggested[92] that Cl^--bridged inner-sphere activated complexes are formed in the halide-catalyzed reaction. The reason for this suggestion was that the ratio of the rate constants of $Fe^{+2}(aq) + Fe^{+3}(aq)$ and $Fe^{+2}(aq) + FeCl^{+2}$ reactions was approximately the same as the ratio for the $Cr^{+2}(aq) + Fe^{+3}(aq)$ and $Cr^{+2}(aq) + FeCl^{+2}$ reactions. In the latter reaction, inert $CrCl^{+2}$ was formed in high yield, strongly indicating that Cr–Cl–Fe bridges were present in the activated complex. It has now been shown[68] that, at 25°C, Cl^- forms a Fe–Cl–Cr bridged inner-sphere activated complex when $Fe(H_2O)_5Cl^{+2}$ and $Cr^{+2}(aq)$ are the reactants and that Cl^- plays a non-bridging role in activated complexes such as $[(H_2O)_5Fe(H_2O) \cdots ClCr(H_2O)_5]^{+4}$ and $[(H_2O)_5Fe—H_2O—CrCl(H_2O)_4]^{+4}$ when $Fe^{+3}(aq)$ and $Cr(H_2O)_5Cl^+$, or $Fe^{+3}(aq)$, Cl^-, and $Cr^{+2}(aq)$, are reactants. The rate constants for the reaction of $Cr^{+2}(aq)$ with $Fe^{+3}(aq)$, $FeOH^{+2}$, and $FeCl^{+2}$ are 2.3×10^3, 3.3×10^6 and 2×10^7 M^{-1} sec.$^{-1}$ (see Ref. 68), respectively, whereas those for the reaction of $Fe^{+2}(aq)$ with the same oxidants are 4, 2980,[61] and 32.6 M^{-1} sec.$^{-1}$ (see Ref. 79), respectively. It is seen that the effect of Cl^- in the former set is much more pronounced than it is in the latter set. Halpern[17] suggested that the small dependence of the rate constant of the $Fe^{+2}(aq) + FeX^{+2}$ reaction on the nature of the halide ion meant that this reaction involved an outer-sphere activated complex. For example, the rate constants of the $Cr(bipy)_3^{+2} + Co^{III}(NH_3)_5X$ reactions (X = H_2O, OH^-, Cl^-), which proceed by outer-sphere mechanisms,[93,94] are only slightly dependent on the nature of X, whereas the rate constants of the $Cr^{+2}(aq) + Co^{III}(NH_3)_5X$ reactions,[21,93,95] which proceed by bridged inner-sphere mechanisms, are much more dependent on the nature of X.

However, the lack of a large X^- effect in $Fe^{+2}(aq) + FeX^{+2}$ is not completely convincing evidence that X^- plays a non-bridging role. It is to be noted that a non-bridging Cl^- in the $Cr^{+2}(aq) + Cl^- + Fe^{+3}(aq)$ reaction also has a very pronounced effect on the rate constant. The third-order rate constant for the latter reaction[68] is 2.2×10^4 M^{-2} sec.$^{-1}$. If the reaction occurs between $CrCl^+$ and $Fe^{+3}(aq)$, then we have $kK = 2.2 \times 10^4$ M^{-2} sec.$^{-1}$, where k is the second-order rate constant for the reaction $CrCl^+ + Fe^{+3}(aq)$ and K is the formation constant for $CrCl^+$. Since the formation constant for $CrCl^{+2}$ is $\sim$0.25 M^{-1} (see Ref. 96), K for $CrCl^+$ is probably less than this. Hence k has a lower limit of approximately 10^5 M^{-1} sec.$^{-1}$ and may even be considerably greater or comparable to the rate constant for the Cl^--bridged inner-sphere mechanism. Indeed it is now known[69] that Cl^- plays both bridging and non-bridging roles in the activated complexes for Cl^--catalyzed ferrous-ferric isotope exchange, and it has been assumed that the Cl^- enters in the $FeCl^{+2}$ complex. In the F^--catalyzed path it has been shown[67] that F^- enters in the FeF^{+2} complex, but it is not presently known whether F^- plays bridging, non-bridging, or both roles.

The case for the other anions (with the notable exception of N_3^- where there is good evidence for a bridged inner-sphere mechanism) which catalyze the ferrous-ferric isotope exchange is less clear. At present one can only speculate concerning the mechanisms of electron transfer for these anions.

Net hydrogen-atom transfer, suggested[32] for

$$Fe(H_2O)_6^{+2} + {}^*Fe(H_2O)_5OH^{+2} \rightleftharpoons$$

$$\left\{ \begin{array}{c} (H_2O)_5Fe\!-\!O \cdots H \cdots O\!-\!{}^*Fe(H_2O)_5 \\ \mid \qquad\qquad\quad \mid \\ H \qquad\qquad\quad H \end{array} \right\}^{\ddagger} \rightleftharpoons$$

$$Fe(H_2O)_5OH^{+2} + {}^*Fe(H_2O)_6^{+2} \quad (3\text{-}9)$$

can also be classed in the bridged outer-sphere mechanism, since the intact inner coordination spheres of the reactants are joined by hydrogen bonds. When the activated complex breaks up, the hydrogen atom can accompany either reactant but it will probably choose to form a bond with that species which will minimize the free energy. (See the discussion following Eq. 3–13.) In reaction (3–9), the hydrogen atom is assumed to move so as to form $FeOH^{+2} + {}^*Fe^{+2}(aq)$ rather than $Fe^{+3}(aq) + {}^*FeOH^+$ since the former products have a lower free energy than the latter. As a result, a net transfer of a hydrogen atom occurs and, as long as both the electron and proton are transferred during the lifetime of the activated complex, the transfers are considered to be simultaneous.

Although net hydrogen-atom transfer is an attractive mechanism for reaction (3–9) it is not particularly attractive for the uncatalyzed reaction (3–8). A net transfer of one hydrogen atom in the latter reaction results in the intermediate formation of $\{(H_2O)_5FeOH^{+2} \cdots (H_3O) \ast Fe(H_2O)_5^{+3}\}$ with energies approaching that for self-ionization of H_2O.[17] However, electron transfer in a symmetrical hydrogen-bridged outer-sphere activated complex can be postulated. The hydrogen bridge consists of a hydrogen bond (although as many as four hydrogen bonds have been postulated[29]) which dissociates without net hydrogen-atom transfer when the activated complex dissociates into products. It must be emphasized that this mechanism should *not* be called "hydrogen-atom transfer" but should be called an "outer-sphere activated complex mechanism." The presence of hydrogen bonding between the intact first coordination shells somewhat stabilizes the activated complex.

Horne and Axelrod[34] have reviewed the evidence for hydrogen transfer along a chain of water molecules by a mechanism similar to that for proton transfer in proton conduction. It was pointed out that the activation energies for electrical conduction in ice and in pure water and for many electron-transfer reactions are approximately equal. This mechanism can explain the electron transfer between Fe(II) and Fe(III) over great distances in ice-like media but it is not popular for the liquid phase because coulombic repulsion between reactants in aqueous media is so small that reactants can readily approach to form an outer-sphere activated complex without a prohibitively large activation energy. Although hydrogen transfer can readily account for the decrease of rate constant with decrease of water content in certain non-aqueous solvents, other reasons can be advanced to explain the experimental facts, as has been pointed out in the section on effect of solvent. Presence of water is not necessary for exchange in all non-aqueous solvents. The ferrous-ferric isotope exchange has a rate constant equal to[85] 16 M^{-1} sec.$^{-1}$ at 25° in DMSO which is independent of acid or water concentrations over small ranges and suggests that electron transfer, involving DMSO complexes in inner- or outer-sphere activated complexes, can occur without hydrogen atom transfer.

OTHER Fe(II) + Fe(III) ELECTRON-TRANSFER REACTIONS

Fe Exchange Between Fe(EDTA)$^{-2}$ and Fe(EDTA)$^{-}$

Here (EDTA)$^{-4}$ is a frequently used symbol for the quadrivalent anion of ethylenediaminetetraacetic acid. The rate of exchange between the two complexes, in the presence of $Fe^{+2}(aq)$, was too rapid to measure by conventional techniques involving separation times of the order of 15 sec.[97] The exchange appears to occur between the versenate complexes or between acidic or basic forms of these complexes according to the reaction

$$Fe(EDTA)^{-2} + \ast Fe(EDTA)^{-} \rightleftharpoons Fe(EDTA)^{-} + \ast Fe(EDTA)^{-2}$$

Exchange does not occur via dissociation of the complexes followed by electron transfer between the aquo ions and reassociation of the complexes. The dissociation constants of these complexes are sufficiently small so that exchange between the aquo ions would be too slow to account for the observed rate. Also, Jones and Long[98] investigated the exchange

$$Fe^{+3}(aq) + *Fe(EDTA)^- = *Fe^{+3}(aq) + Fe(EDTA)^-$$

and found that the exchange proceeded at a measurable rate, so that dissociation of $Fe(EDTA)^-$ could not have been involved in the mechanism by which the rapid exchange between the EDTA complexes of Fe(II) and Fe(III) occurred.

Likewise, the exchange does not occur via reactions (3–27) and (3–28):

$$Fe^{+2}(aq) + *Fe(EDTA)^- = Fe^{+3}(aq) + *Fe(EDTA)^{-2} \qquad (3\text{--}27)$$

$$Fe^{+2}(aq) + *Fe(EDTA)^{-2} = *Fe^{+2}(aq) + Fe(EDTA)^{-2} \qquad (3\text{--}28)$$

Jones and Long[98] found reaction (3–28) too rapid to measure, but Reynolds, Liu, and Mickus[86] found that reaction (3–27) was too slow to play a part in the rapid exchange between the EDTA complexes of Fe(II) and Fe(III). This mechanism was considered by Reynolds, Liu, and Mickus because the standard free-energy change of reaction (3–27) is approximately 14.8 kcal./mole so that a second-order rate constant for the forward reaction might have been as large as 100 l./mole sec., i.e., sufficiently large to have permitted the exchange observed by Adamson and Vorres.[97]

Fe Exchange Between 1,10-Phenanthroline Complexes of Fe(II) and Fe(III)

Attempts to measure the rate of isotope exchange between $Fe(ph)_3^{+2}$ and $Fe(ph)_3^{+3}$ (see Refs. 89, 76) and between the tris-5,6-dimethylphenanthroline iron(II) and iron(III) complexes[99] have been unsuccessful, although a lower limit of 10^5 l./mole sec. has been established for the former exchange.

The rate constants for dissociation of $Fe(ph)_3^{+2}$ and $Fe(ph)_3^{+3}$ have been determined.[100] For $Fe(ph)_3^{+2}$, $k = 4.5 \times 10^{-3}$/min. at 25°C in the pH range 0.3 to 2.3; for $Fe(ph)_3^{+3}$, $k = 3.1 \times 10^{-3}$/min. at 25°C in 1 M H_2SO_4. Hence the mechanism of electron transfer does not involve dissociation of either complex prior to exchange. The most probable mechanism is electron transfer in an outer-sphere activated complex. The electron transferred is probably from a metal-to-ligand $d\pi$ bond which is delocalized over the metal ion and all three phenanthroline ligands. Intercomplex transfer is thought to occur through overlap of the π systems of the two complexes. The overlap of two π molecular orbitals, one on $Fe(ph)_3^{+2}$ and one on $Fe(ph)_3^{+3}$, has been calculated by Mataga and Reynolds[101] at a distance of approach determined by bond lengths, bond angles, and the van der Waals radius of the hydrogen atoms attached to the rings. Although the overlap cannot be obtained accurately, it is certainly sufficiently large to allow easy transfer of the electron, provided reorganization of the ligand and solvent spheres has been accomplished. Since the exchange is very rapid, the reorganization process requires only a small $\Delta F^{\ddagger}$. By the method of crossed reactions (see Chapter 6), Dulz and Sutin[102] calculated a value of 2×10^3 l./mole sec. at 25°C for the rate constant, but this value is lower than the established lower limit.

Sutin and co-workers[87,88,90] have investigated various reactions between $Fe^{+2}(aq)$ and dipyridyl, tripyridyl, phenanthroline, and substituted phenanthroline complexes of Fe(III) and between various Fe(II) and Fe(III) dipyridyl and phenanthroline complexes, some of which are listed in Table 3–1. The rate constants of these reactions are usually very large, some being greater than 10^8 l./mole sec. and a few being as small as 10^3 to 10^4 l./mole sec. Those with rate constants greater than 10^8 l./mole sec. are probably diffusion-controlled. The $Fe^{+2}(aq) + Fe(ph)_3^{+3}$ reaction is interesting since $E_{act} = 0.8$ kcal./mole and $\Delta S^{\ddagger} = -37$ cal./deg. mole. It was suggested[90] that the large negative entropy change was due to ferrous ion penetrating the space between the phenanthroline rings in order to approach the ferric ion for reaction. However, this value is an apparent entropy change obtained from the absolute-rate-theory expression

$$k = \frac{eRT}{Nh} \exp\left(\frac{\Delta S^{\ddagger}}{R}\right) \exp\left(-\frac{E_{act}}{RT}\right)$$

so as to give agreement between measured values of the rate constant k and the activation energy E_{act}. The apparent entropy of activation will include contributions from a transmission coefficient if the latter is less than unity in addition to true entropy changes accompanying the formation of the activated-complex configuration from the separated reactants. We shall inquire into the possibility of forming an outer-sphere activated complex between $Fe(H_2O)_6^{+2}$ and $Fe(ph)_3^{+3}$, in which it will be assumed that electron tunnelling is a slow step, using the minimum reorganization of inner and outer coordination spheres necessary to achieve energy balance in the activated complex upon electron transfer.

The activation process will be considered to consist of several steps, each with its accompanying free-energy change. The sum of the free energy changes is the free energy of activation. The steps are:

1. Reorganization of inner coordination and solvation shells of the separated reactants to the radii which they have in the activated complex. The enthalpy and free-energy changes are given by Eqs. 2-27 and 2-28.
2. Approach of the two reorganized reactants to a distance where their inner coordination spheres just touch.
3. Reorganization of the solvent shells at the constant inner-shell radii achieved in step 1.

It is assumed that the steps are approximately independent of each other and that electrostatic theory can be used to get an estimate of the energy change in each step.

The over-all free-energy change ΔF°, calculated from the oxidation potentials of $-0.77\,V$ and $-1.06\,V$ for Fe^{+2}/Fe^{+3} and $Fe(ph)_3^{+2}/Fe(ph)_3^{+3}$, respectively, is -6.7 kcal./mole. In terms of the activation free energies of the forward and reverse reactions, ΔF° is

$$\Delta F^\circ = \Delta F_f^{\ddagger} - \Delta F_b^{\ddagger}$$

where $\Delta F_f^{\ddagger} = w_i^R + w_r^R + w_0^R$ and the w's are the free-energy changes associated with steps (1), (2), and (3), respectively, for the reactants. A similar expression may be written for $\Delta F_b^{\ddagger}$ in terms of analogous w's for the products. Thus we have the following equation which determines the value of $r_{\ddagger}$ in the activated complex:

$$\Delta F^\circ = (w_i^R - w_i^P) + (w_0^R - w_0^P) + (w_r^R - w_r^P) = -6.7 \text{ kcal.}$$

In this reaction $w_r^R = w_r^P$. For the calculation of a minimum $\Delta F^{\ddagger}$, it is assumed that reorganization of the inner spheres of $Fe(ph)_3^{+2}$ and $Fe(ph)_3^{+3}$ is unnecessary; this assumption has support from the fact that, since the isotope exchange reaction between the two complex ions is very rapid, the free energy of activation which is equal to or greater than the inner-sphere reorganization free energy is small. Hence $(w_i^R - w_i^P)$ involves only reorganization energies from $Fe(H_2O)_6^{+2}$ and $Fe(H_2O)_6^{+3}$. The work of reorganizing $Fe(H_2O)_6^{+2}$ is given by

$$w_i^R = U^{\ddagger}(+2) - U^0(+2) - \frac{4e^2}{2}\left(1 - \frac{1}{\varepsilon}\right)\left\{\frac{r_0 - r_{\ddagger}}{(r_0 + r_{H_2O})(r_{\ddagger} + r_{H_2O})}\right\}$$

and the first term in Eq. 2–28 has been dropped. $U^{\ddagger}(+2)$ is the potential energy of $Fe(H_2O)_6^{+2}$ with the ion-dipole distance, $r_{\ddagger}$, characteristic of the activated complex. $U^0(+2)$ is the potential energy of $Fe(H_2O)_6^{+2}$ in the ground state. Actually, U^0 should be corrected for the zero-point energy, but this approximately cancels in the difference $(w_i^R - w_i^P)$. A similar expression holds for w_i^P.

The reorganization energies of the outer solvent shells are calculated from Marcus' equations (see Chapter 6) for m, λ_0, and $\lambda = \lambda_0 + \lambda_i$. Here $\lambda_i = (U^{\ddagger} - U^0)/m^2$ since $m^2\lambda_i$ is the reorganization energy of the inner coordination shell. A value of m is obtained by a successive-approximations method after starting with m for both reactants and products equal to $(-\frac{1}{2})$. Only two cycles are necessary to get constant values of w_i^R, w_i^P, w_0^R and w_0^P. It was found that $r_{\ddagger} = 2.13$ Å, $w_0^R = 5.2$ kcal., $w_i^R = 4.8$ kcal. when corrected for a zero-point energy of approximately 1 kcal.[16] and $w_r^R = 0.5$ kcal. Thus the activation free energy of the forward reaction is 10.5 kcal./mole compared with the observed value of 11.3 kcal./mole. The agreement between the calculated and observed $\Delta F^{\ddagger}$ values indicates that a contribution of $-RT \ln \kappa$ from the transmission coefficient is negligible. However, the apparent agreement is misleading because $\Delta H^{\ddagger}$ makes a major contribution to the calculated $\Delta F^{\ddagger}$ but an almost negligible contribution to the experimental $\Delta F^{\ddagger}$.

Fe Exchange Between $Fe(CN)_6^{-4}$ and $Fe(CN)_6^{-3}$

A number of the early investigators found the rate of this exchange too rapid to measure,[103,104,105] but Wahl and co-workers[91,106] have succeeded in determining rate laws for it. The rate is first-order in $Fe(CN)_6^{-4}$ and $Fe(CN)_6^{-3}$ concentrations, but no path of exchange appears to be independent of cation concentration. When tetraphenylarsonium ion, Ph_4As^+, is used in the presence of EDTA to complex all cationic impurities, the rate is given by[106]

$$R = k[Fe(CN)_6^{-4}][Fe(CN)_6^{-3}][Ph_4As^+]$$

with $k = 2.8 \times 10^3$ l.^{2}/mole^2 sec.$^{-1}$ at 0.1°C, and is independent of ionic strength and the nature of the anions present. In KOH media, the rate is given by an expression similar to that derived for anionic catalysis of the $Fe^{+2}(aq) + Fe^{+3}(aq)$ exchange:

$$R = \left\{ \frac{k_1 K_1[K^+] + k_2 K_1 K_2[K^+]^2}{1 + K_1[K^+] + K_1 K_2[K^+]^2} \right\} [Fe(CN)_6^{-4}][Fe(CN)_6^{-3}]$$

At 0.1°C, $k_1 = 180$ l./mole sec., $k_2 = 1.1 \times 10^5$ l./mole sec., $K_1 = 60$ l./mole and $K_2 = 0.6$ l./mole.

Since the inner coordination spheres of the ferrocyanide and ferricyanide ions are inert to substitution, it is probable that electron transfer occurs in a bridged outer-sphere activated complex. In view of the $d\pi$ bonding which can occur between the metal ions and the ligands, with consequent delocalization of the t_{2g} electrons, it is surprising that direct electron transfer in an activated complex containing only $Fe(CN)_6^{-4}$ and $Fe(CN)_6^{-3}$ is not a predominant path. According to Marcus' theory for the uncatalyzed path, $\Delta F^{\ddagger} = 10.1$ kcal./mole at zero ionic strength if one assumes that no inner coordination sphere reorganization is necessary,[7] whereas $\Delta F_1^{\ddagger}$ (corresponding to k_1 for the first K^+ catalyzed path) is equal to 14.4 kcal./mole. In view of the fact that the inner coordination sphere reorganization energy of $Fe(H_2O)_6^{+2}$ (in the reaction with $Fe(ph)_3^{+3}$) is calculated to be only 4 kcal./mole, the inner-sphere reorganization energy should not be much greater than this for $Fe(CN)_6^{-4}$. Indeed, the inner-sphere reorganization energy of $Fe(CN)_6^{-4}$ in this isotope exchange reaction is usually considered to be less than for $Fe(H_2O)_6^{+2}$ so that the total free energy of activation would not exceed 14.4 kcal./mole. The small negative value of ΔF° in the $Fe(H_2O)_6^{+2} + Fe(ph)_3^{+3}$ reaction helps to lower $\Delta F^{\ddagger}$ for the heteronuclear electron-transfer reaction as compared to the ferrocyanide–ferricyanide isotope exchange but the effect is small, not over-riding, because $|m| = 0.45$ for the former reaction and 0.50 for the latter (see Chapter 6). Why is the uncatalyzed reaction so slow ($\Delta F^{\ddagger}$ considerably in excess of 14.4 kcal./mole)? Is the inner-sphere rearrangement free energy large? This has not been thought to be the case, since the two ions do not have markedly different shapes or sizes. The rearrangement free energy of the outer solvation spheres is approximately 5 kcal./mole (see Chapter 6 for calculation of outer sphere rearrangement energies) and hence is not excessive. Furthermore, whether or not 5 kcal./mole is an accurate value for this quantity, it is likely that it has very similar values in both the K^+-catalyzed and in the uncatalyzed reactions, so outer-sphere rearrangement cannot be responsible for the difference in rates. If the rate of the uncatalyzed reaction is slow because the transmission coefficient is much less than unity, then this reaction is unique in having the only outer-sphere activated complex with $\kappa \ll 1$. If this proved to be true in the future, it may be related to the fact that the unpaired $d\pi$ electron in $Mo(CN)_8^{-3}$ is not spread over the nitrogen nuclei although it reaches the carbon nuclei of the CN^- ligands.[107]

THE Np(IV) + Np(VI) REACTION

Very many electron-transfer reactions involving ions of the actinide series have been studied. Most of these reactions have features of interest, so that a proper selection of examples is difficult. The Np(IV) + Np(VI) reaction and its reverse, the disproportionation of Np(V), and the Np(IV) + Np(V) isotope exchange reaction have been chosen because of their related nature.

The Forward Reaction

The equilibrium between NpO_2^+ and $Np^{+4} + NpO_2^{+2}$ at pH 0 is in favor of formation of Np(V), so the reaction between Np(IV) and Np(VI) was studied[108,109] most conveniently by following the decrease of Np^{+4} absorption with time. The rate of the forward reaction between Np(IV) and Np(VI) in ClO_4^- media was found[108] to be given by

$$R_f = k_f \frac{[\text{Np(IV)}][\text{Np(VI)}]}{[\text{H}^+]^2} = k_{obs}[\text{Np(IV)}][\text{Np(VI)}] \qquad (3\text{–}29)$$

The values of the activation parameters $E_{a,f}$, $\Delta H_f^{\ddagger}$, and $\Delta S_f^{\ddagger}$ at $\mu = 2.0$ were 25.2 ± 1.6 kcal./mole, 24.6 ± 1.6 kcal./mole, and 17.8 ± 2.7 cal./deg. mole, respectively, so that at 25°C, $\Delta F_f^{\ddagger} = 19.3 \pm 1.6$ kcal./mole. In the second, more careful study,[110] two pH-dependent paths were found, so that the rate was given by

$$R_f = \left[\frac{k_{1f}}{[\text{H}^+]^2} + \frac{k_{2f}}{[\text{H}^+]^3} \right] [\text{Np(IV)}][\text{Np(VI)}] \qquad (3\text{–}30)$$

where $k_{2f} \cong 0.1 k_{1f}$. Since most experiments were carried out with $[H^+] \geqslant 1$ M, the second term can be neglected in most of the discussion. In this study, the apparent activation energy of the observed second-order rate constant was 25.32 ± 0.28 kcal./mole. The average dependence on hydrogen ion concentration was given by $[H^+]^{-2.14}$.

The rate-determining step of the path involving $[H^+]^{-2}$ was apparently Eq. 3–31 or Eq. 3–32.

$$Np(OH)_2^{+2} + NpO_2^{+2} \xrightarrow{k_{31}} 2NpO_2H^{+2} \qquad (3\text{–}31)$$

$$NpO^{+2} + NpO_2^{+2} \xrightarrow{k_{32}} NpO_2^+ + NpO^{+3} \qquad (3\text{–}32)$$

If the reaction was Eq. 3–31, then the observed rate constant, k_{obs}, at a pH where the second path is unimportant, is given by

$$k_{obs} = \frac{k_{31} K_{2h}}{[H^+]^2 + K_{1h}[H^+] + K_{2h}} \cong \frac{k_f}{[H^+]^2} \qquad (3\text{–}33)$$

in which K_{1h} and K_{2h} are the first and second over-all hydrolysis constants of Np^{+4}. In order that k_{obs} be inversely proportional to $[H^+]^2$, the conditions

$$K_{1h} \ll [H^+] \qquad \text{and} \qquad K_{2h} \ll [H^+]^2 \qquad (3\text{–}34)$$

must be satisfied. Hence k_f will be given by

$$k_f = k_{31} K_{2h} \qquad (3\text{–}35)$$

and the activation parameters given above for k_f contain contributions from K_{2h}. Before the thermodynamic quantities for the activation reaction of the electron-transfer step can be evaluated, K_{2h} must be studied as a function of temperature. Of course, if the electron-transfer reaction was Eq. 3–32, then analogous arguments would lead to the equality $k_f = k_{32} K_{2h}'$, where K_{2h}' is the equilibrium constant for the formation of NpO^{+2} from Np^{+4} and H_2O. A lower limit for k_{31} (or k_{32}) may be established, since k_f is known and K_{2h} (or K_{2h}') has an upper limit imposed by the condition in Eq. 3–34. Thus, for $1.00\ M\ H^+$, $k_f = 4.7 \times 10^{-2}$ mole/l. sec., $K_{2h} \leqslant 0.1$ and $k_{31} \geqslant 0.5$ l./mole sec.

Transition States

The electron-transfer reaction between Np(IV) and Np(VI) may be formulated either as hydrogen-atom transfer, as an electron tunnelling mechanism or as an oxo-bridged inner-sphere activated-complex mechanism. The hydrogen-atom transfer for the first path may be represented by

$$Np(OH)_2^{+2} + NpO_2^{+2} \rightleftharpoons$$
$$\left\{ HO\text{—}Np^{IV}\text{—}O \cdots H \cdots O\text{—}Np^{VI}\text{—}O \right\}^{\ddagger} \rightarrow 2NpO_2H^{+2} \qquad (3\text{–}36)$$

$$NpO^{+2} + NpO_2^{+2} \rightleftharpoons$$
$$\left\{ \begin{array}{c} O\text{—}Np^{IV}\text{—}O \cdots H \cdots O\text{—}Np^{VI}\text{—}O \\ | \\ H \end{array} \right\}^{\ddagger} \rightarrow 2NpO_2H^{+2} \qquad (3\text{–}37)$$

or by some variation of the central idea expressed in these reactions. In Eq. 3–36, $Np(OH)_2^{+2}$ is the reacting Np(IV) species; in Eq. 3–37, NpO^{+2} is the reacting Np(IV) species. In Eq. 3–37, a water molecule in the solvation shell of NpO^{+2} is involved in the hydrogen-atom transfer. The two transition states differ only in the position of one hydrogen atom, and because they are so similar only one will be considered.

The oxo-bridged transition-state mechanism may be represented by

$$NpO^{+2} + {}^*NpO_2^{+2} \rightleftharpoons \{O{-}Np^{IV} \cdots O \cdots {}^*Np^{VI}{-}O\}^{\ddagger} \rightarrow NpO_2^+ + {}^*NpO^{+3}$$

$$(3\text{–}38)$$

The oxygen atom can remain equally well with either neptunium ion when the transition state dissociates to products.

A considerable number of transition states for electron tunnelling may be written down; a few typical ones will be given. When $Np(OH)_2^{+2}$ is assumed to be a reactant, one may write

$$Np(OH)_2^{+2} + NpO_2^{+2} \rightleftharpoons \left\{ \begin{array}{c} O \cdot | \cdot H \cdots O \\ | \quad\quad\; | \\ Np^{IV} \rightsquigarrow^{e^-} Np^{VI} \\ | \quad\quad\;\; | \\ O \cdots H \cdot | \cdot O \end{array} \right\}^{\ddagger} \rightarrow 2NpO_2H^{+2} \quad (3\text{–}39)$$

$$(I)$$

or, if water molecules in the solvation shells are employed to hydrogen bond the reacting species in the transition state, reactions such as

$$Np(OH)_2^{+2} + NpO_2^{+2} \rightleftharpoons \left\{ \begin{array}{c} H \\ | \\ O \\ \cdots \quad / \\ H \quad / \quad H \\ \backslash \quad / \quad \cdot \\ O \cdot \quad\quad O \\ | \quad\quad\;\; | \\ Np^{IV} \rightsquigarrow^{e^-} Np^{VI} \\ | \quad\quad\;\; | \\ O \quad\quad\quad O \\ \backslash \\ H \end{array} \right\}^{\ddagger} \rightarrow 2NpO_2H^{+2} \quad (3\text{–}40)$$

$$(II)$$

The hydrogen bond in I should not be interpreted as being linear. After electron tunnelling, the transition state breaks apart at the bonds marked by dashed lines. It is to be noted that the transition states I and II could also be used in hydrogen-atom transfer mechanisms. When NpO^{+2} is assumed to be a reactant, one might write

$$NpO^{+2} + *NpO_2^{+2} \rightleftharpoons \{O—Np \cdots O \cdots *Np—O\}^{\ddagger} \rightarrow NpO_2^+ + *NpO^{+3}$$

(III)

(or)

$$\rightarrow NpO^{+3} + *NpO_2^+ \quad (3\text{-}41)$$

or, if water molecules in the hydration shells are employed to hydrogen bond the reacting species in the transition state, reactions such as

$$NpO^{+2} + NpO_2^{+2} \rightleftharpoons$$

$$\left(\left\{ O—Np^{IV}—O \quad\quad O—Np^{VI}—O \right\} \begin{matrix} H \\ \\ \\ H \quad e^- \end{matrix} \right)^{\ddagger} \rightarrow 2NpO_2H^{+2} \quad (3\text{-}42)$$

(IV)

Transition states I and IV both have two hydrogen bonds and differ only in their configurations; they could be distinguished only if it were known whether $Np(OH)_2^{+2}$ or NpO^{+2} was a reactant. Because of the close approach of Np nuclei in transition states such as I and III, it is likely that the potential barrier for tunnelling is sufficiently small so that tunnelling is not a slow step in these symmetrical transition states. Once the activation energy has been acquired by the reacting species and they have formed the symmetrical transition state I or III, the electron will be readily distributed between the two Np nuclei. The slow step of the reaction would be formation of the symmetrical transition state from the reactants. If this is true of Eqs. 3–39 and 3–41, then a discussion of the mechanism of Eq. 3–39 is identical to a discussion of hydrogen-atom transfer having I as transition state, and a discussion of the mechanism of Eq. 3–41 is identical to a discussion of Eq. 3–38.

When the Np nuclei approach less closely, as in Eqs. 3–40 or 3–42, electron tunnelling may be a slow step. In this case, formation of a favorable or symmetrical configuration in the transition state is accompanied only rarely by transition of an electron from an orbital involving Np(IV) to one involving Np(VI). The small tunnelling probability for the electron now governs the rate of charge transfer within the transition state.

The $NpOH^{+3} + NpO_2^{+2}$ Reaction

An interesting point to note is that $NpOH^{+3}$ does not contribute appreciably to the reaction between Np(IV) and NpO_2^{+2}. The hydrolysis constant for Np^{+4} is 5.0×10^{-3} M (see Ref. 111), so that hydrolysis of Np^{+4} to form $NpOH^{+3}$ is as important as hydrolysis of Fe^{+3} to form $FeOH^{+2}$.

If $NpOH^{+3}$ was to react with NpO_2^{+2} by a bridged transition state similar to that of Eq. 3–38, then the mechanism postulated for the forward reaction between Np^{+4} and NpO_2^{+2} would probably be

$$Np^{+4} + H_2O \rightleftharpoons NpOH^{+3} + H^+ \text{ (rapid)} \qquad K_{1h} \qquad (3\text{–}43)$$

$$NpOH^{+3} + {}^*NpO_2^{+2} \rightleftharpoons \{HO\text{—}Np^{IV} \cdots O^- \cdots {}^*Np^{VI}\text{—}O\}^{\ddagger}$$
$$\rightarrow NpO_2H^{+2} + {}^*NpO^{+3} \quad (3\text{–}44)$$

In the absence of evidence to the contrary, it would be assumed that the equilibria

$$NpO_2^+ + 2\,H^+ \rightleftharpoons NpO^{+3} + H_2O \qquad K_{45} \qquad (3\text{–}45)$$

$$NpO_2^+ + H^+ \rightleftharpoons NpO_2H^{+2} \qquad\qquad\qquad (3\text{–}46)$$

are rapidly established. The equilibrium in Eq. 3–46 is not likely to be slowly established, since it involves only an ionization of a proton.

If it is assumed for purposes of discussion that the rate-determining step of the actual reaction *is* Eq. 3–38, then it is difficult to understand why the forward-rate constant in Eq. 3–38 should be so much larger than that in Eq. 3–44. The rate of the forward reaction via Eq. 3–38 is given by

$$R_{38} = k_{38,f}K'_{2h}\frac{[Np^{+4}][NpO_2^{+2}]}{[H^+]^2} = k_{0,f}[Np^{+4}][NpO_2^{+2}]$$

where $k_{38,f}$ is the second-order forward-rate constant for Eq. 3–38 and $k_{0,f}$ is the observed forward-rate constant given by

$$k_{0,f} = k_{38,f}K'_{2h}/[H^+]^2$$

Similarly, the rate of the forward reaction via Eq. 3–44 is given by

$$R_{44} = k_{44,f}K_{1h}\frac{[Np^{+4}][NpO_2^{+2}]}{[H^+]} = k'_{0,f}[Np^{+4}][NpO_2^{+2}]$$

where $k_{44,f}$ is the second-order forward-rate constant of Eq. 3–44, K_{1h} is the equilibrium constant for Eq. 3–43, and $k'_{0,f}$ is the observed forward rate constant given by

$$k'_{0,f} = k_{44,f}K_{1h}/[H^+]$$

The rate of the $Np^{+4} + NpO_2^{+2}$ reaction was studied over the concentration range given by $0.227 \leqslant [H^+] \leqslant 1.99\,M$. At pH 0, $k_{0,f} \gg k'_{0,f}$ since dependence on $NpOH^{+3}$ was not observed.[110] Since $K_{2h} < K_{1h}$, this result implies $k_{38,f} \gg k_{44,f}$. The inner-sphere bridge mechanism of Eq. 3–38 does not readily explain this inequality.

If the rate-determining step is hydrogen-atom transfer, where the hydrogen atom transferred is the one in $NpOH^{+3}$, as in

$$NpOH^{+3} + {}^*NpO_2^{+2} \rightleftharpoons \{Np^{IV}\text{—}O \cdots H \cdots O\text{—} {}^*Np^{VI}\text{—}O\}^{\ddagger} \rightarrow$$
$$\xrightarrow{\quad\quad} NpO^{+3} + {}^*NpO_2H^{+2} \quad (3\text{–}47)$$

the small rate of Eq. 3–47 as compared to Eq. 3–36 may be explained if NpO^{+3} in Eq. 3–47 is formed with difficulty as compared to NpO_2H^{+2} in Eq. 3–36. If the hydrogen atom transferred belongs to a water molecule in the hydration shell, then the reaction for $NpOH^{+3}$ may be written as in

$$NpOH^{+3} + {}^*NpO_2^{+2} \rightleftharpoons$$

$$\left(\begin{array}{c} HO-Np^{IV} \cdots O \cdots H \cdots O-{}^*Np^{VI}-O \\ \mid \qquad \longrightarrow \\ H \end{array} \right)^{\ddagger} \longrightarrow$$

$$Np(OH)_2^{+3} + {}^*NpO_2H^{+2} \quad (3\text{–}48)$$

As the hydrogen atom is transferred toward $^*NpO_2^{+2}$, a second OH group becomes bonded to the original $Np(IV)$ so that the atomic configuration about the resulting $Np(V)$ approaches that of the final product, NpO_2H^{+2} (or NpO_2^+), much more closely. However, the intermediate $Np(OH)_2^{+3}$ is formed. If the equilibrium constant for the formation of this ion is too small, then the forward and reverse reactions in Eq. 3–48 would contribute negligibly toward establishing equilibrium between Np^{+4}, NpO_2^{+2}, and NpO_2^+. Assumption of hydrogen-atom transfer mechanism makes the difference between the rates of reaction of $NpOH^{+3}$ and $Np(OH)_2^{+2}$ with NpO_2^{+2} more readily explained. A study of the acid-base equilibria in solutions of NpO_2^+ is indicated.

If the mechanism of electron transfer is tunnelling in transition states similar to II in Eq. 3–40 or IV in Eq. 3–42, then the slow rate of a step involving $NpOH^{+3}$, as compared to that involving $Np(OH)_2^{+2}$ (or NpO^{+2}) may be explained on the lack of a second O atom in $NpOH^{+3}$ which may be easily positioned so that tunnelling of an electron from $Np(IV)$ to $Np(VI)$ can occur.

The Reverse Reaction

The equilibrium constant for the over-all reaction,

$$Np^{+4} + NpO_2^{+2} + 2\,H_2O \rightleftharpoons 2NpO_2^+ + 4H^+ \qquad K_{49} \qquad (3\text{–}49)$$

at 25°C and $\mu = 1.00$ is 5.45×10^6 $(mole/l.)^4$ in 1 M $HClO_4$,[111] and the changes in the thermodynamic functions are[112]

$$\Delta F_{49} = -9.19 \text{ kcal./mole}$$

$$\Delta H_{49} = +7.5 \text{ kcal./mole}$$

$$\Delta S_{49} = 56 \text{ cal./deg. mole}$$

Combining these values with the values for the forward reaction (at $\mu = 2.00$), the approximate values for the reverse (disproportionation) reaction may be calculated from $k_b = k_f/K_{49}$ and are

$$\Delta F_b^{\ddagger} = 28.5 \pm 1.6 \text{ kcal./mole}$$
$$\Delta H_b^{\ddagger} = 17.1 \pm 1.6 \text{ kcal./mole} \qquad (3\text{-}50)$$
$$\Delta S_b^{\ddagger} = -38.1 \pm 2.9 \text{ cal./deg. mole}$$

The values in Eq. 3–50 for the disproportionation reaction of Np(V) will be compared with values from the Np(IV) + Np(V) exchange. From the concentration dependencies of the paths for the forward reaction and of the equilibrium, Eq. 3–49, the concentration dependence of the disproportionation reaction should be given by

$$R_{\text{dis}} = k_{1b}[\text{NpO}_2^+]^2[\text{H}^+]^2 + k_{2b}[\text{NpO}_2^+]^2[\text{H}^+] \qquad (3\text{-}51)$$

The observed rate constant of the forward reaction (see Eq. 3–30) is given by[109]

$$k_{\text{obs},f} = \frac{k_{1f}}{[\text{H}^+]^2} + \frac{k_{2f}}{[\text{H}^+]^3} = \frac{k_f}{[\text{H}^+]^{2.14}} \qquad (3\text{-}52)$$

over the pH range studied, and that of the reverse reaction should be given by

$$k_{\text{obs},b} = k_{1b}[\text{H}^+]^2 + k_{2b}[\text{H}^+] = k_b[\text{H}^+]^{1.86} \qquad (3\text{-}53)$$

over the same pH range. This hydrogen ion dependence, derived from measurements on the Np(IV) + Np(VI) reaction, will be compared with that obtained from a study of the disproportionation reaction encountered in the Np(IV) + Np(V) isotope exchange.

THE Np(IV) + Np(V) EXCHANGE

An opportunity of studying the Np(V) disproportionation reaction occurred in the Np(IV) + Np(V) isotope exchange reaction. Besides possible direct electron transfer between the (IV) and (V) valence states, isotope exchange could occur through the reverse and forward reactions of the equilibrium in Eq. 3–49. The rate of exchange was stated[113] to be given by

$$R = k_1[\text{NpO}_2^+]^2[\text{H}^+] + \frac{k_2[\text{Np}^{+4}]^{3/2}[\text{NpO}_2^+]^{1/2}}{[\text{H}^+]^2} \qquad (3\text{-}54)$$

As was evident from the discussion preceding Eq. 3–53, the first term of R is expected to be $k_b[\text{NpO}_2^+]^2[\text{H}^+]^{1.86}$ because the pH dependence of the first term was studied within the pH range employed in the study of the Np(IV) + Np(VI) reaction. If the total rate of isotope exchange, R, in Eq. 3–54, is written as

$$R = R_H + R_L \qquad (3\text{-}55)$$

in which R_H and R_L are the rates of the "high-acid" and "low-acid" paths, respectively, then a re-evaluation of the data[112] gives*

$$R_H = k_1[H^+]^2[NpO_2]^2$$

The High-Acid Path ($0 \leqslant$ pH $\leqslant 0.35$)

The following mechanisms may be postulated for the exchange at high-acid concentration:[113]

Mechanism A

$$*NpO_2^+ + H^+ \rightleftharpoons *NpO_2H^{+2} \text{ (rapid)} \qquad K_{46} \qquad (3-46)$$

$$*NpO_2H^{+2} + NpO_2H^{+2} \rightleftharpoons X_I^{\ddagger} \xrightarrow{k_{56}} *Np(OH)_2^{+2} + NpO_2^{+2} \text{ (slow)} \quad (3-56)$$

$$*Np^{+4} + 2H_2O \rightleftharpoons *Np(OH)_2^{+2} + 2H^+ \text{ (rapid)} \qquad K_{2h} \qquad (3-57)$$

Mechanism B

$$NpO_2^+ + 2H^+ \rightleftharpoons NpO^{+3} + H_2O \text{ (rapid)} \qquad K_{45} \qquad (3-45)$$

$$NpO^{+3} + *NpO_2^+ \rightleftharpoons X_{II}^{\ddagger} \xrightarrow{k_{58}} *NpO^{+2} + NpO_2^{+2} \text{ (slow)} \qquad (3-58)$$

$$*Np^{+4} + H_2O \rightleftharpoons *NpO^{+2} + 2H^+ \text{ (rapid)} \qquad K'_{2h} \qquad (3-59)$$

The disproportionation reaction is rate-determining in each mechanism. In mechanism A, the rate of exchange is given by

$$R_H = k_{56}K_{46}^2[NpO_2^+]^2[H^+]^2 = k_1[NpO_2^+]^2[H^+]^2 \qquad (3-60)$$

where k_1 is the observed fourth-order rate constant. A similar expression involving $k_{58}K_{45}$ is derived for mechanism B. Equation 3–60 is to be compared to Eq. 3–51. If the second term in Eq. 3–51 is neglected because the observed order of reaction with respect to hydrogen ion concentration is approximately 2, then k_1 should be equal to k_{1b}. The value of k_1 (obtained from Tables II and III of Sullivan, Cohen, and Hindman[113]) at 1.00 M $HClO_4$ is $(9.25 \pm 0.70) \times 10^{-5}$ l.^{3}/mole^3 sec. at 47°C and 1.2×10^{-5} l.^{3}/mole^3 sec. at 25°C. The value of k_{1b} is obtained from k_{1f} of Eq. 3–52 and the equilibrium constant of Eq. 3–49. At 25°C, $k_{1f} = 4.27 \times 10^{-2}$ mole/l. sec.,[109] $K = 5.45 \times 10^6$ (mole/l.)^4 (see Ref. 112), so that $k_{1b} = 7.85 \times 10^{-9}$ l.^{3}/mole^3 sec. Although some of the values used here pertain to different ionic strengths, this source of error is not serious. The rate constants k_1 and k_{1b} differ by a factor of approximately 1.5×10^3. This is difficult to understand. If the disproportionation path represented by k_1 is so much faster than the

---

* The values of R in the pH range 0 to 0.4 must be corrected for the contributions from R_L. At pH = 0.00, 0.10, 0.24, 0.28, 0.33, $R = 2.37, 1.89, 1.32, 1.21, 1.15 \times 10^{-6}$ M/min., respectively; $R_L = 1.34, 2.12, 4.04, 4.85, 6.11 \times 10^{-7}$ M/min., respectively; and $R_H = 2.24, 1.68, 0.92, 0.72, 0.54 \times 10^{-6}$ M/min. When log R_H is plotted against log [H$^+$], the slope of the line is approximately 2.

one represented by k_{1b}, then there should be a corresponding forward-rate constant for $Np^{+4} + NpO_2^{+2}$ which is much larger than k_{1f}.

The activation parameters[113] for k_1 at 25° and $\mu = 1.2$ are given in Table 3-2. The activation parameters of k_{1b} from Eq. 3-50 are also given in Table 3-2, for comparison. The difference between k_1 and k_{1b} is reflected in $\Delta F_1^{\ddagger}$ and $\Delta F_b^{\ddagger}$. Use of significance tests on the differences $\Delta F_b^{\ddagger} - \Delta F_1^{\ddagger}$ and $\Delta H_b^{\ddagger} - \Delta H_1^{\ddagger}$ show that the former may possibly be significant but that the latter is definitely not significant. Since the $\Delta F^{\ddagger}$ difference has not been proven significant and since the $\Delta H^{\ddagger}$'s are not significantly different, it

TABLE 3-2

High-Acid Exchange	Reverse Reaction of Eq. 3-49
$\Delta F_1^{\ddagger} = 24.2 \pm 0.2$ kcal./mole	$\Delta F_b^{\ddagger} = 28.5 \pm 1.6$ kcal./mole
$\Delta H_1^{\ddagger} = 17.6 \pm 2.3$ kcal./mole	$\Delta H_b^{\ddagger} = 17.1 \pm 1.6$ kcal./mole
$\Delta S_1^{\ddagger} = -22.2 \pm 7.2$ cal./deg. mole	$\Delta S_b^{\ddagger} = -38 \pm 2.7$ cal./deg. mole

cannot be stated with certainty whether the high-acid path for the Np(IV) + Np(V) exchange and the reverse of the $Np(OH)_2^{+2} + NpO_2^{+2}$ reaction are the same or different.

The Low-Acid Path $(0.5 \leqslant pH \leqslant 1.4)$

The following mechanism may be postulated to explain the observed concentration dependence:[113]

$$Np^{+4} + {}^*NpO_2^+ \rightleftharpoons Np^{+3} + {}^*NpO_2^{+2} \text{ (rapid)} \quad K_{61} \qquad (3\text{-}61)$$

$$Np^{+4} + 2 H_2O \rightleftharpoons Np(OH)_2^{+2} + 2 H^+ \text{ (rapid)} \quad K_{2h} \qquad (3\text{-}57)$$

or

$$Np^{+4} + H_2O \rightleftharpoons NpO^{+2} + 2 H^+ \text{ (rapid)} \quad K_{2h}' \qquad (3\text{-}59)$$

$$Np(OH)_2^{+2} \text{ (or } NpO^{+2}) + {}^*NpO_2^{+2} \rightleftharpoons X_{III}^{\ddagger} \xrightarrow{k_{62}}$$
$$\qquad\qquad {}^*Np(OH)_2^{+2} \text{ (or } {}^*NpO^{+2}) + NpO_2^{+2} \text{ (slow)} \qquad (3\text{-}62)$$

The rate of exchange by this path is given by

$$R_L = k_{62}[Np(OH)_2^{+2}][NpO_2^{+2}]$$

$$= k_{62}K_{2h}K_{61}^{1/2} \frac{[Np^{+4}]^{3/2}[NpO_2^+]^{1/2}}{[H^+]^2} \qquad (3\text{-}63)$$

upon substitution for $[NpO_2^{+2}]$ from

$$[NpO_2^{+2}] = [Np^{+3}] = (K_{61}[Np^{+4}][NpO_2^+])^{1/2} \qquad (3\text{-}64)$$

The observed rate constant, k_2 (see Eq. 3-54), is given by

$$k_2 = k_{62}K_{2h}\sqrt{K_{61}} \qquad (3\text{-}65)$$

Of course, if NpO^{+2} were assumed to be the reactant in place of $Np(OH)_2^{+2}$ then K'_{2h} would appear in place of K_{2h}.

It is of interest to enquire whether the transition state formed from $Np(OH)_2^{+2}$ and NpO_2^{+2} in Eq. 3–62 is the same one formed in the reproportionation reaction, Eq. 3–31. From Eq. 3–35, we have $k_f \simeq k_{1f} = k_{31}K_{2h}$, and from Eq. 3–65 we have $k_2/\sqrt{K_{61}} = k_{62}K_{2h}$. If both transition states had the same nuclear configuration, then, except for differences in transmission coefficients, $k_{31} = k_{62}$, and

$$k_f = k_{62}K_{2h} = k_2/\sqrt{K_{61}} \qquad (3\text{–}66)$$

The values of $\Delta F^{\ddagger}$, $\Delta H^{\ddagger}$, and $\Delta S^{\ddagger}$ measured for k_2 and k_f are given in the second and fifth columns of Table 3–3; the values of the thermodynamic

TABLE 3–3

Comparison of Thermodynamic Quantities Derived from the Np(IV) + Np(VI) Reaction with Those Derived from the Low-Acid Path of the Np(IV) + Np(V) Exchange at 25°

	k_2^*	K_{61}^{**}	$k_2/\sqrt{K_{61}}$	k_f^{***}
ΔF (kcal./mole)	27.2 ± 0.2	22.6	15.9 ± 0.2	19.3 ± 1.6
ΔH (kcal./mole)	36.8 ± 2.2	33.8	19.9 ± 2.2	24.6 ± 1.6
ΔS (cal./deg. mole)	32.1 ± 6.9	37.6	13.3 ± 6.9	17.8 ± 2.7

$^*\mu = 1.2$.
$^{**}\mu = 1.0$.
$^{***}\mu = 2.0$.

quantities corresponding to K_{61} are given in the third column and were measured by Cohen and Hindman.[112] Those corresponding to $k_2/\sqrt{K_{61}}$ are given in the fourth column. The values in the fourth and fifth columns are to be compared. The t test for the significance of the difference between the $\Delta H^{\ddagger}$ values shows that the difference is insignificant ($t = 1.5$) although the difference between the $\Delta F^{\ddagger}$ values may be significant ($t = 2.1$). If the difference in $\Delta F^{\ddagger}$ values is significant, it means that the exchange reaction of $Np(OH)_2^{+2}$ and NpO_2^{+2} is $10^2 - 10^4$ times faster than the reproportionation reaction of $Np(OH)_2^{+2}$ and NpO_2^{+2}. If true, this variation in rates is due chiefly to a difference in *frequency factors* since the enthalpies of activation are not significantly different.

Although the results are not all for the same ionic strength, the errors are, in many cases, considerable, so that correction for this factor would not significantly alter the results.

The question as to whether the Np(IV) + Np(VI) = 2Np(V) reaction and the "high-" and "low-acid" exchanges for Np(IV) + Np(V) all involve the same transition state cannot be answered at present but remains an intriguing question. It is interesting to note that

$$\frac{k_2}{k_1\sqrt{K_{61}}} = \exp\left[\frac{1}{RT}(24{,}200 - 15{,}900)\right] = 1.3 \times 10^6 \simeq K = \frac{k_f}{k_b}$$

where K is the equilibrium constant of Eq. 3–49.

It should be noted that in the low-acid exchange the concentrations of Np^{+3} and Np(VI) are equal, so that a postulate of a rate-determining step such as

$$Np(OH)_2^{+2} + {}^{*}Np^{+3} \xrightarrow{k_{67}} Np^{+3} + {}^{*}Np(OH)_2^{+2} \qquad (3\text{–}67)$$

would also lead to the observed concentration dependence. The exchange of Np(III) and Np(IV) should be investigated to determine whether or not the activation free energy, enthalpy, and entropy for $k_{67}K_{2h}$ would be equal to those listed for $k_2/\sqrt{K_{61}}$ in Table 3–3.

It must also be noted that if the equilibrium in Eq. 3–61 was rapidly established, and the evidence is that it is, then the equilibrium would lead to rapid exchange between Np(IV) and Np(V) if the Np atoms were equivalent in the transition state of reaction (3–61). Therefore, if the equilibrium is part of the mechanism of the low-acid path of exchange and since rapid exchange is not observed, it must be concluded that Np atoms are not equivalent in the transition state of reaction Eq. 3–61. Either Np(V) always forms Np(VI) in the forward reaction so that Np(VI) always forms Np(V)—never Np(IV)—in the reverse reaction, or otherwise Np(V) always forms Np(III) in the forward reaction and Np(III) always forms Np(V)—never Np(IV)—in the reverse reaction. The former possibility seems to be much more reasonable, since NpO_2^+ has two correctly positioned oxygen atoms for the formation of NpO_2^{+2}, whereas Np(IV) apparently does not. Most reactions involving oxidation of Np(III) or Np(IV) to Np(V) or Np(VI), or the reverse reductions, are slow, whereas oxidation of Np(III) to Np(IV) and of Np(V) to Np(VI), and the reverse reductions, are rapid.[114] For exchange to occur by the postulated mechanism, NpO_2^+ must always form NpO_2^{+2} in Eq. 3–61 and NpO_2^{+2} must always form NpO_2^+. If NpO_2^+ formed Np(III) and Np(IV) formed NpO_2^{+2}, then exchange between Np(IV) and NpO_2^{+2} in Eq. 3–62, followed by NpO_2^{+2} going to Np(IV) in the reverse reaction in Eq. 3–61, would *not* lead to exchange between Np(IV) and Np(V). However, in this case, exchange between Np(IV) and Np(III) could be postulated, as in Eq. 3–67, and exchange between Np(IV) and Np(V) would have the observed concentration dependencies.

Absence of an [Np(IV)][Np(V)] Term in the Np(IV) + Np(V) Exchange Rate

Lack of a term involving [Np(IV)][Np(V)] in the rate law eliminates the possibility of direct exchange between an Np(IV) species and NpO_2^+. Exchange between the Np(IV) and Np(V) oxidation states is accomplished by indirect processes. In the high-acid path, disproportionation of Np(V) forms Np(IV) and NpO_2^{+2}. The Np(IV) is then mixed with the rest of the Np(IV) in solution. Formation of Np(V) occurs by the reaction of the mixed Np(IV) with NpO_2^{+2}. The result is exchange between Np(IV) and Np(V). The rate constant, k_{31}, for the $Np(OH)_2^{+2} + NpO_2^{+2}$ reproportionation reaction is large compared to the rate constant for the disproportionation of NpO_2^+ because the equilibrium in $HClO_4$ media is much in favor of Np(V) formation. On the contrary, the rate constant for the $Np(IV) + NpO_2^+$ reaction is small compared to the rate constant for the disproportionation reaction of NpO_2^+ because exchange between Np(IV) and NpO_2^+ occurs by the latter reaction rather than the former at high-acid concentrations. One reason why k_{31} for the one-electron transfer between Np(IV) and NpO_2^{+2} is greater than that for the one-electron transfer reaction between Np(IV) and NpO_2^+ is that in the former reaction $\Delta F° \simeq -9k$ cal./mole whereas in the latter $\Delta F° \simeq 0$ kcal./mole. In the theory of Marcus[9] (see Chapter 6) a relationship was derived between the over-all free-energy change, $\Delta F°$, and the free energy of activation, $\Delta F^{\ddagger}$, which would lead one to expect that the former reaction would be the more rapid.

In the low-acid path for exchange, reaction between Np(IV) and *NpO_2^+ to give Np(III) and *NpO_2^{+2} first occurs. Then an Np(IV) species reacts with *NpO_2^{+2} by a two-electron transfer process to produce *Np(IV) and NpO_2^{+2}. The exchanged NpO_2^{+2} species can now react with Np(III) in the reverse reaction of the rapidly established equilibrium between Np(IV), NpO_2^+ and Np(III) and NpO_2^{+2}. The result is exchange between Np(IV) and Np(V). The rate constant, k_{62}, for the $Np(OH)_2^{+2} + NpO_2^{+2}$ exchange reaction is much larger than that for the $Np(OH)_2^{+2} + NpO_2^+$ exchange reaction because the former reaction leads to exchange between Np(IV) and Np(V) at low-acid concentrations instead of the latter reaction, despite the small concentration of NpO_2^{+2} as compared to NpO_2^+. Why? In both of these exchange reactions, presumably $\Delta F° = 0$. Also, electron transfer occurs between Np^{+4} and NpO_2^+ to give Np^{+3} and NpO_2^{+2} with $\Delta F° = 22.6$ kcal./mole (see Table 3–4), but not directly between Np^{+4} and *NpO_2^+ to give *Np^{+4} and NpO_2^+ with $\Delta F° = 0$ kcal./mole. Why?

The reaction between Np^{+3} and NpO_2^+ to form $2Np^{+4}$ is rapid at 25°C, although here the rate can probably be ascribed to a negative $\Delta F°$ value of approximately -13 kcal./mole at this temperature.

Transition States

The transition state formed from the two reactants $Np(OH)_2^{+2} + NpO_2^{+2}$ should explain these facts:

1. One electron (reproportionation) or two electrons (exchange) can be transferred within nearly identical transition states (if the $\Delta F^{\ddagger}$ values of 15.9 and 19.3 kcal./mole in Table 3-3 are not significantly different).

2. NpO_2^{+2} in the transition state cannot be replaced by NpO_2^+, although the two ions are virtually identical[115] except for charge, since the rate constant for exchange between $Np(OH)_2^{+2}$ and NpO_2^{+2} is so much greater than that for exchange between $Np(OH)_2^{+2}$ and NpO_2^+.

3. $Np(OH)_2^{+2}$ in the transition state cannot be replaced by $Np(OH)^{+3}$ with very small effect. (An example of a very small effect is in the replacement of $FeCl_2^+$ by $FeCl^{+2}$ as oxidant for $Fe^{+2}(aq)$.)

These requirements would appear to eliminate a transition state such as that in Eq. 3–38 since there would seem to be no reason why the analogous one-electron reaction

$$NpO^{+2} + {}^*NpO_2^+ \rightleftharpoons \{O-Np^{IV} \cdots O \cdots {}^*Np^V-O\}^{\ddagger} \rightarrow NpO_2^+ + {}^*NpO^{+2}$$

$$(3-68)$$

could not occur with a rate comparable to or faster than that of Eq. 3–38.

The transition state I in Eq. 3–39 would appear to be suitable for Np(IV) + Np(V) exchange as in

$$Np(OH)_2^{+2} + {}^*NpO_2^+ \rightleftharpoons$$

$$\left(\begin{array}{c} O \cdot | \cdot H \cdots O \\ | \quad | \quad {}_{e^-} \quad | \\ Np^{IV} \rightsquigarrow {}^*Np^V \\ | \quad | \quad \quad | \\ O \cdot | \cdot H \cdots O \\ | \end{array} \right)^{\ddagger} \rightarrow NpO_2^+ + {}^*Np(OH)_2^{+2} \quad (3-69)$$

In fact, hydrogen-bond formation ought to occur more easily in Eq. 3–69 than in Eq. 3–39 so that the Np(IV) + Np(V) direct exchange should occur as readily as the reproportionation reaction between Np(IV) and NpO_2^{+2}.

The same objection can be raised to transition state IV in Eq. 3–42. It would seem that the hydrogen bonds formed in IV would be more easily formed by NpO_2^+ than by NpO_2^{+2} and that Np(IV) + Np(V) exchange could occur by

$$NpO^{+2} + {}^*NpO_2^+ \rightleftharpoons$$

$$\left(\begin{array}{c} | H \\ | \cdot \\ O-Np^{IV}-O \cdot | \quad \cdot O-{}^*Np^V-O \\ | \cdot \\ | H \quad {}_{e^-} \end{array} \right)^{\ddagger} \rightarrow NpO_2^+ + {}^*NpO^{+2} \quad (3-70)$$

Since the exchange does not proceed by a path of this nature which is comparable in rate to the Np(IV) + Np(VI) reproportionation reaction, transition state IV is probably not formed in the latter reaction.

The hydrogen-atom transfer mechanisms, Eqs. 3–36 and 3–37, provide more satisfactory explanations of some of the observations. To bring about exchange between Np(IV) and Np(V) by a transfer of one hydrogen atom, one may write equations such as

$$Np(OH)_2^{+2} + {}^*NpO_2^+ \rightleftharpoons$$

$$\left\{HO-Np^{IV}-O \cdots H \xrightarrow{\quad} O-{}^*Np^V-O\right\}^{\ddagger} \rightarrow$$

$$NpO_2H^{+2} + {}^*NpO_2H^+ \quad (3\text{–}71)$$

$$NpO^{+2} + {}^*NpO_2^+ \rightleftharpoons$$

$$\left\{\begin{matrix} O-Np^{IV}-O \cdots H \xrightarrow{\quad} O-{}^*Np^V-O \\ \quad\quad | \\ \quad\quad H \end{matrix}\right\}^{\ddagger} \rightarrow$$

$$NpO_2H^{+2} + {}^*NpO_2H^+ \quad (3\text{–}72)$$

These equations involve the Np(IV) species NpO_2H^+, whose existence is doubtful. The reverse reactions of Eqs. 3–71 and 3–72 would probably contribute extremely little to equilibrium in the exchange and hence the forward reactions would contribute negligibly to exchange also. However, Eqs. 3–36 or 3–37 provide reasonable paths for the Np(IV) + Np(VI) reproportionation reaction. It has already been shown (page 54) that Eq. 3–36 or Eq. 3–37 can be used to explain why $NpOH^{+3}$ reacts slowly with NpO_2^{+2} in comparison to $Np(OH)_2^{+2}$. However, neither reaction (3–36) nor (3–37) explains how two-electron transfer processes can occur at least as frequently as one-electron transfer in the same transition state. Perhaps the exchange between Np(IV) and Np(VI) occurs by a different transition state than reproportionation, i.e., the $\Delta F^{\ddagger}$ values in Table 3–3 for $k_2/\sqrt{K_{61}}$ and k_f are significantly different.

CHROMIUM AND COBALT OXIDATION-REDUCTION REACTIONS

As a result of the informative studies of Taube and his co-workers, some reactions of chromium and cobalt have become among the best-known of modern inorganic chemistry. The reactions are of the type

$$Co(NH_3)_5X^{+3-z} + Cr^{+2}(aq) \rightleftharpoons \{[(NH_3)_5Co \cdots X \cdots Cr(H_2O)_5]^{+5-z}\}^{\ddagger} \rightarrow$$

$$5NH_3 + Co^{+2}(aq) + CrX^{+3-z} \quad (3\text{–}73)$$

$$Co(NH_3)_6^{+3} + Cr(bipy)_3^{+2} \rightleftharpoons \{[(NH_3)_5Co(NH_3)(bipy)Cr(bipy)_2]^{+5}\}^{\ddagger} \rightarrow$$

$$6NH_3 + Co^{+2}(aq) + Cr(bipy)_3^{+3} \quad (3\text{–}74)$$

and were used to demonstrate inner- and outer-sphere activated complexes, respectively.

$Co(NH_3)_5X^{+3-z} + Cr^{+2}(aq)$ Reactions

For ease of writing, the $Co(NH_3)_5^{+3}$ group will be designated R and the charge on the ion will be omitted except where necessary to prevent ambiguity. Thus the Co(III) ion in the heading becomes RX.

The first reactions which were postulated to proceed by the bridged inner-sphere activated complex mechanism involved $Cr^{+2}(aq)$ and RX with $X = F^-$, Cl^-, Br^-, I^-, SO_4^{-2} (see Ref. 21), N_3^-, $P_2O_7^{-4}$, acetate, n-butyrate, crotonate, succinate, oxalate, maleate,[116] SCN^- (see Ref. 116, 117), H_2O, OH^- (see Refs. 118, 93), fumarate, mono-esters of maleic and fumaric acids, isomeric forms of phthalic acid,[22,23,24,119,120] and malonate.[121] This list is not meant to be exhaustive, but typical, of the range of X groups which were found to be transferred from the inert reactant complex ion, RX, to the inert Cr(III)—X product complex ion. The transfer of X groups is a very strong indication that the X group is present in the inner coordination spheres of both metal ions in the activated complex, i.e., that electron transfer proceeds through an inner-sphere activated complex.

Transfer of Cl^- Ion

In the absence of free Cl^- in solution, Cl^- was quantitatively transferred from RX to $CrCl^{+2}$ (see Refs. 20, 21). Labelled free Cl^- added to the solution, but not to the reactant RX, appeared in the product $CrCl^{+2}$ only to a very minor extent.[20] Hence it was concluded that transfer of Cl^- in Eq. 3–73 occurs without release of Cl^- to the solution at any stage, followed by re-entry into the first coordination sphere of either Co or Cr. The presence of free pyrophosphate ion in an $RX + Cr^{+2}(aq)$ reaction mixture resulted in both Cl^- and $P_2O_7^{-4}$ being bonded to Cr(III) in the product.[116,122] Apparently the pyrophosphate ion entered the activated complex in the inner co-ordination sphere of Cr(II) and was bonded to Cr(III) upon electron transfer from Cr(II) to Co(III) in a Cl^--bridged inner-sphere activated complex. The pyrophosphate ion plays a non-bridging role in the activated complex. This experiment was a clear demonstration that anions can be present in the activated complex and not be required for bridging.

A lower limit of 2×10^6 l./mole sec.[95] has been established for the assumed second-order rate constants of the halide (F^-, Cl^-, Br^-, I^-) ion transfer reactions.

Transfer of NCS^-

The transfer of thiocyanate ion was studied[117] with the intention of determining whether $Cr—SCN^{+2}$ or $Cr—NCS^{+2}$ was formed from R—NCS. If $Cr—SCN^{+2}$ had been found, it would have strongly indicated that the bridged activated complex was Co—NCS—Cr with the metal ions attached to either end of the thiocyanate ion. The results were inconclusive. The product Cr(III)-thiocyanate complex had a spectrum identical to that of

$(H_2O)_5Cr$—NCS^{+2}. However, experiments with radioactively labelled thio-cyanate ion showed that the thiocyanate ion was at least partially released to the solution during transfer so that isomerization from Cr—SCN^{+2} to Cr—NCS^{+2} could have occurred during this time.

The enthalpy and entropy of activation were[95] 6.9 kcal./mole and -29 cal./deg. mole, respectively.

Transfer of H_2O and OH^-

Transfer of ^{18}O from ROH_2 and from ROH to the product Cr^{+3}(aq) occurred.[118,123] The observed rate constant was of the form[118,80]

$$k_{obs} = k_0 + \frac{k_h K_h}{[H^+]}$$

in which k_0 is the rate constant for $ROH_2 + Cr^{+2}$(aq) and k_h is that for $ROH + Cr^{+2}$(aq). The rate constants at 20°C, $\mu = 1.2$, were[80] 0.5 l./mole sec. and 1.7×10^6 l./mole sec., respectively, and $\Delta H_0^{\ddagger} = 2.9$ kcal./mole, $\Delta S_0^{\ddagger} = -52$ cal./deg. mole, $\Delta H_h^{\ddagger} = 4.6$ kcal./mole, $\Delta S_h^{\ddagger} = -18$ cal./deg. mole, the subscript identifying the electron-transfer path to which the activation parameter belongs. The values of K_h and ΔH_h employed in the calculation of k_h and $\Delta H_h^{\ddagger}$ were[80] 9.3×10^{-7} M and 9.5 kcal./mole, respectively.

Fractionation of ^{18}O and ^{15}N resulted[118] in the enrichment of the Co(III) complex in $H_2{}^{18}O$ and in $^{15}NH_3$ as the reaction progressed. Both results are consistent with an activated complex in which water is the bridging group and the Co—NH_3 distance has been increased to a value lying between its values in the Co(II) and Co(III) complexes.

The small $\Delta H_0^{\ddagger}$ and very negative $\Delta S_0^{\ddagger}$ values recall a similar situation in the Fe^{+2}(aq) + $Fe(ph)_3^{+3}$ reaction. It is difficult to find sources in such small ions for a decrease of 52 entropy units. It will be recalled that cal-culations for $Fe(H_2O)_6^{+3}$ in Chapter 2 showed that dielectric saturation probably did not extend into the second hydration shell, so that the ion would have little influence on solvent molecules beyond the second hydration shell. Hence it seems that a part of $\Delta S_0^{\ddagger}$ is due to a transmission coefficient of less than unity. The reaction is probably spin-forbidden, due to the Co(III) and Co(II) complex ions being spin-paired and spin-free, respectively.

In the presence of free pyrophosphate, sulfate, maleate[116] and Cl^- ions[118] in solution, the two former ions were found to be complexed with Cr(III) in the product whereas the latter two were complexed to a negligible extent. Rates of reaction were not measured, so the extent to which the non-bridging pyrophosphate and sulfate ions catalyze the electron transfer through the H_2O- or OH^--bridged activated complex is unknown at present.

With the oxidant cis-$Co(NH_3)_4(H_2O)_2^{+3}$ or cis-$Co(en)_2(H_2O)_2^{+3}$ the trans-fer of oxygen to Cr(III) is complete and only one oxygen is transferred for each $Cr(H_2O)_6^{+3}$ formed.[123] At 25°C and 1.0 M H^+, the rate constant· for the Cr^{+2}(aq) + Cr^{+3}(aq) isotope exchange reaction is approximately

0.5 l./mole hr.[124] and is too small to have permitted appreciable loss of $H_2{}^{18}O$ from the first coordination sphere of Cr(III) during the time of the Co(III) + Cr^{+2}(aq) reaction.

Transfer of Organic Ligands

Many organic ligands have been used and the number is growing apace because, in this case, subtle variations can be built into the bridging group to observe the effect on electron transfer.[125,126,127,128,129]

The organic ligands may be divided into two broad categories, namely, monofunctional and polyfunctional ligands. Both categories are joined to the sixth coordination position of $Co(NH_3)_5$ through a functional group. In the case of monofunctional ligands this is the only functional group available; in the case of polyfunctional ligands the other functional groups are left free. Reaction between RX and a reductant such as Cr^{+2}(aq) is usually classified as "adjacent attack" or "remote attack." [116] Adjacent attack occurs when Co(III) and the reductant are coordinated to the same functional group in the activated complex, electron transfer presumably occurring through this functional group. Remote attack occurs when Co(III) and the reductant are coordinated to different functional groups and a net electron transfer occurs through a system of orbitals of proper symmetry involving the functional groups and that portion of the ligand which lies between the functional groups. For example, among the many Co(III) complexes for which adjacent attack by Cr^{+2}(aq) has been postulated are the acetato, n-butryrato, crotonato, succinato, methyl succinato,[116] benzoato, p-hydroxybenzoato,[128] and isophthalato[119] complexes since the second-order rate constants were of the same order of magnitude and since the organic ligand was bonded to Cr(III) in the products and Co(III) was reduced to Co^{+2}(aq). Although a second functional group is present in some of these ligands, it is not used by Cr^{+2}(aq) in the activated complex and a common mechanism of electron transfer through one carboxylate group exists. In the succinate ligand, the two carboxyl groups are joined by a saturated carbon chain and the failure of an electron to be readily transferred through this is easily understood. In the crotonate ligand, $CH_3CH{=}CHCO_2^-$, Cr^{+2}(aq) apparently does not attack at the double bond. In the p-hydroxy-benzoate and isophthalate ligands, a complete system of conjugated double bonds between the two metal ions does not exist for the remote attack mechanism. However, the rate of reaction is not entirely independent of the organic group attached to the one functional group.[128]

Remote attack* has been postulated[116] for ligands such as oxalate, maleate, fumarate, methyl monoesters of maleic and fumaric acids,[116,22,23,24] tereph-

---

* Taube indicated at the 151st meeting of the American Chemical Society in Pittsburgh, Pa., March 22–31, 1966, that the picture is not as clear as it was and that the remote-attack mechanism is under further study.

thalate[119] and p-aldehydobenzoate.[130]   In all of these ligands a conjugated double-bond system joins the two metal ions attached to different functional groups.   This conjugated double-bond system is utilized in the net transfer of one electron from Cr(II) to Co(III).   The organic ligand is coordinated with Cr(III) in the products and Co(III) is reduced to $Co^{+2}(aq)$.   Evidence for remote attack may be summarized as follows:

1. The rate constants for the reactions where remote attack is probable are frequently much larger than those for adjacent attack.   If both metal ions were attached to the same carboxyl group of oxalate ligand, there would not be a good reason why the rate constant should be so much larger than those for the acetato, butyrato, crotonato, succinato, etc., complexes.

2. The monomethylmaleate and monomethylfumarate ligands were hydrolyzed[22] during the transfer from Co(III) complex to Cr(III) complex. The methyl alcohol formed by the hydrolysis and the organic acid were both coordinated to the product Cr(III).   When the reactions were carried out in enriched $^{18}OH_2$ (see Ref. 23) it was found that 97–98 percent of the methanol derived its oxygen from the water.   This water was bound in the first coordination shell of Cr(III), or the methanol would not have been coordinated to Cr(III).   The monomethylsuccinate ligand did not hydrolyze in the reaction between the Co(III) complex and $Cr^{+2}(aq)$.   Remote attack by $Cr^{+2}(aq)$ on the esterified carboxyl groups of the two unsaturated acids and adjacent attack on the carboxyl group of the saturated acid can most conveniently explain these facts.

3. The maleate and methylmaleate ligands (*cis* form) were partly converted to fumarate ligand (*trans* form) during the electron transfer reaction.[24,131] The fraction converted was pH-dependent and increased with increasing H^+ concentration.   Use of $V^{+2}(aq)$ in place of $Cr^{+2}(aq)$ gave similar results.   When the electron-transfer reaction with $V^{+2}(aq)$ was carried out in D_2O, the fumaric acid produced contained C—D bonds, whereas the maleic acid did not.   It was concluded[24,131] that a new C—H bond (on an unsaturated C atom) was formed when the bridge ligand "absorbed" an electron from the reducing ion, isomerization occurred, and that a C—H bond was broken as the bridging ligand transferred its electron to the oxidizing ion.   A relatively long-lived metastable state of the ligand was apparently formed.   When the maleate bridging ligand did not isomerize apparently the ligand did not retain an unpaired electron for a sufficient length of time to either add H^+ in the formation of new C—H bonds or to isomerize without the addition of H^+.   Exchange of —CH=CH— hydrogen atoms with deuterium atoms in D_2O or isomerization to maleic acid did not occur when the fumarato complex of Co(III) was used.

4. Remote attack occurs only when delocalized molecular orbitals (e.g., those for conjugated double-bond systems) of sufficiently low energy are available to interact with the pertinent metal atomic orbitals and to join the metal atomic orbitals through the MO.   Relatively strong σ bonds can be formed between the metal ions and carboxyl groups but, as in the case of dibasic succinic acid, this alone does not result in electron transfer.

Since the early work with organic ligands established the main mechanism of adjacent and remote attack, the more recent investigations have been concerned with learning more detail about these mechanisms. Variations of the remote functional group,[130] of substituents on a given ligand,[128] of ligand,[127,128] of ligands to replace NH_3 in the R group,[132] and of reducing ion[130,95] have been made.

The inner-sphere mechanisms are not always exclusively adjacent attack or remote attack; sometimes the two mechanisms compete.

Reaction Between Cr^{+2}(aq) and Co(III) Amine Complexes

The rates of reaction of Cr^{+2}(aq) with $Co(NH_3)_6^{+3}$ and $Co(en)_3^{+3}$ are very slow compared to the rates of reaction with RX oxidants. At 25°C the rate constants are 9×10^{-5} (see Ref. 93) and 2×10^{-5} l./mole sec.,[95] respectively, and $\Delta H^{\ddagger} = 14.7$ kcal./mole, $\Delta S^{\ddagger} = -30$ cal./deg. mole for the former.

The reaction with RNH_3 is catalyzed by Cl^- to a very marked extent; the rate constant at 25°C is 1.2×10^{-2} l.^{2}/mole^2 sec. with $\Delta H^{\ddagger} = 12.4$ kcal./mole and $\Delta S^{\ddagger} = -25$ cal./deg. mole. $CrCl^{+2}$ is a product of the catalyzed path. Since RNH_3^{+3} is inert to substitution a Cl^--bridged inner-sphere mechanism is improbable, so that Cl^- exerts its marked influence on the reaction rate while in a non-bridging, inner-coordination-sphere position. This result is markedly different from the effect of Cl^- on the RH_2O^{+3} reaction.

An outer-sphere mechanism is assumed[95] for the RNH_3, catalyzed and uncatalyzed, and for the $Co(en)_3^{+3}$ reactions. The over-all $\Delta F°$ of reaction (3–75) is approximately -12 kcal./mole:

$$Cr^{+2}(aq) + Co(NH_3)_6^{+3} \rightarrow Cr^{+3}(aq) + Co(NH_3)_6^{+2} \qquad (3-75)$$

(dissociation to Co^{+2}(aq) and $6NH_4^+$ will give rise to a further free-energy decrease) and is more negative than the $\Delta F°$ values of very rapid reactions such as the $Fe^{+2}(aq) + Fe(ph)_3^{+3}$ reaction. Consequently, the $\Delta F°$ value is not a decisive factor here. If rearrangement of solvent outside of the first coordination spheres is all that is involved in forming the activated complex, then, according to the Marcus-Hush theory,[6] reaction (3–75) should have a rate constant comparable to, or greater than, that for the iron reaction. Since the $\Delta S^{\ddagger}$ values for the two reactions are comparable, it is a difference in $\Delta H^{\ddagger}$ values which is responsible for the difference in rate constants. Differences in first-coordination-sphere rearrangement energies will be the main contributors the $\Delta H^{\ddagger}$ difference. There will be a greater loss of LFSE in forming the activated complex of reaction (3–75) than in the iron reaction because it is thought that, in the latter, $Fe(ph)_3^{+2}$ and $Fe(ph)_3^{+3}$ have the same nuclear configurations within the amplitude of the ground vibrational state and the change of LFSE for $Fe(H_2O)_6^{+2}$ is small and negative (see estimate in

Chapter 2). It is to be noted here that the rate constant of

$$Cr(bipy)_3^{+2} + Co(NH_3)_6^{+3} + 6\ H^+ \rightarrow Cr(bipy)_3^{+3} + Co^{+2}(aq) + 6NH_4^+ \quad (3\text{--}76)$$

is equal to[133] 6.9×10^2 l./mole sec. at 25°, a factor of 8×10^6 greater than that for $Cr^{+2}(aq)$ in place of $Cr(bipy)_3^{+2}$. Reaction (3–76) is assumed to have an outer-sphere mechanism. Apparently rearrangement energy for the inner coordination sphere of $Co(NH_3)_6^{+3}$ is not a serious energy barrier to the addition of an electron in reaction (3–76). Although requirements for a reactant in one activated complex cannot be carried over unchanged for the same reactant in another activated complex, it is tempting to suppose that rearrangement energy for the inner coordination sphere of $Co(NH_3)_6^{+3}$ cannot be a serious energy barrier to addition of an electron to Co(III) in reaction (3–75) either. In this case, the inner-sphere rearrangement of $Cr^{+2}(aq)$ makes a very important contribution to the activation energy of reaction (3–75). The inner-sphere rearrangement energy of $Cr^{+2}(aq)$ may be especially large because of the tetragonal distortion which is suspected of being present in this ion.

Reactions Between RX and Other Reductants

The most frequently used substitutes for $Cr^{+2}(aq)$ have been $V^{+2}(aq)$, $Eu^{+2}(aq)$, $Cr(bipy)_3^{+2}$, and $Co(CN)_5^{-3}$. $V^{+2}(aq)$ and $Eu^{+2}(aq)$ gave complete hydrolysis of the methyl fumarate ligand,[23] the same as $Cr^{+2}(aq)$, but Fe(II), $Cr(bipy)_3^{+2}$, and $V(bipy)_3^{+2}$ did not.

Zwickel and Taube[93] compared the rates of reaction of $Cr^{+2}(aq)$, $V^{+2}(aq)$, and $Cr(bipy)_3^{+2}$ with RX (X = NH_3, H_2O, OH^-, Cl^-) and concluded that $V^{+2}(aq)$ reactions occurred by the outer-sphere mechanism. The basis for this conclusion was that the variation of V^{+2} rate constants with changing X groups more closely resembled the $Cr(bipy)_3^{+2}$ rate-constant variation than the Cr^{+2} rate-constant variation. The former reacted by outer-sphere mechanisms and the latter by bridged inner-sphere mechanisms (with the possible exception of X = NH_3). Candlin, Halpern, and Trimm[95] extended these comparisons to include $Eu^{+2}(aq)$ and many more Co(III) complexes. The conclusion was that variation of X did indeed produce very similar variations of the $V^{+2}(aq)$ and $Cr(bipy)_3^{+2}$ rate constants. The variation of the $Eu^{+2}(aq)$ rate constants was more difficult to analyze and it was suggested[95] that some, at least, went by an inner-sphere mechanism.

Reduction of RX with $Co(CN)_5^{-2}$ showed that some reductions involved inner-sphere mechanisms, whereas others involved outer-sphere mechanisms.[135] Inner-sphere mechanisms were observed for X = Cl^-, N_3^-, NCS^-, and OH^-. Outer-sphere mechanisms were observed for X = NH_3, SO_4^{-2}, acetate$^-$, fumarate^{-2}, oxalate^{-2}, maleate^{-2}, succinate^{-2}, CO_3^{-2}, and PO_4^{-3}. Both mechanisms were observed for F^-.

Endicott and Taube[94] found that $Ru(NH_3)_6^{+2}$, which is substitution-inert and reacts by an outer-sphere activated-complex mechanism, is similar to $V^{+2}(aq)$ and $Cr(bipy)_3^{+2}$ in behavior rather than $Cr^{+2}(aq)$.

D_2O Effect

The ratio of the rate constants, k_H/k_D, for $RX + Cr^{+2}(aq)$ in H_2O and D_2O was 2.6 for OH^--bridged and 3.8 for H_2O-bridged inner-sphere reactions.[80] The rate constants for the reaction of $Co(NH_3)_5(D_2O)^{+3}$ and $Co(ND_3)_5(D_2O)^{+3}$ with $Cr^{+2}(aq)$ were approximately equal.[80] The latter result was meant to show that H-isotope substitution in non-bridging, inner-shell ligands produced a negligible effect and consequently substitution of D_2O for non-bridging H_2O in $Cr^{+2}(aq)$ would have only a small effect. This assumption was later shown to be approximately correct[93] because the ratio was 1.3 for the $RNH_3^{+3} + Cr^{+2}(aq)$ reaction. Therefore most of the decrease noted in D_2O for hydroxyl- and water-bridged inner-sphere reactions was due to the isotope substitution in the bridge group.

The ratios for $V^{+2}(aq)$ and $Cr(bipy)_3^{+2}$ reacting with RNH_3^{+3} in H_2O and D_2O were 1.7 and 1, respectively. The result for $Cr(bipy)_3^{+2}$ shows that, in this case, the effect from outer solvation shells was very small. The results for $V^{+2}(aq)$ and $Cr^{+2}(aq)$ show that there was an effect from non-bridging, inner-shell water ligands although the effect was smaller for non-bridging water than for bridging water or water ions.

The ratios for $V^{+2}(aq)$ and $Cr(bipy)_3^{+2}$ reacting with ROH_2^{+3} in H_2O and D_2O were both 2.6. These results are somewhat difficult to reconcile with the explanations given above. If the outer solvation shells contribute little to the effect on $ROH_2^{+3} + Cr(bipy)_3^{+2}$, then all of the decrease was due to one inner-shell, but non-bridging, water molecule in ROH_2^{+3}. Also, the effect of the inner-shell, but non-bridging, water molecule in ROH_2^{+3} was much smaller for $V^{+2}(aq)$, which has been assumed to react by an outer-sphere mechanism, than it was for $Cr(bipy)_3^{+2}$. In the $V^{+2}(aq)$ reaction a factor of approximately 1.7 (see result with RNH_3) came from the vanadous ion. However, the D-isotope substitution effects due to inner- and outer-shell, bridging and non-bridging substitutions probably do not remain constant from one reaction to another even for the same reducing ion in the series. For example, the rate ratio for the $RCl^{+2} + V^{+2}(aq)$ reaction, which is also assumed to have an outer-sphere mechanism (or possibly a Cl^--bridged inner-sphere mechanism), was 2.2, a larger ratio than for the $RNH_3^{+3} + V^{+2}(aq)$ reaction. Furthermore, the rate-constant ratio is temperature-dependent so that the ratio observed depends on the temperature, arbitrarily chosen, at which the measurements were made so that the usefulness of this ratio is quite limited.

Cr(II) + Cr(III) Isotope Exchange Reactions

The isotope exchange between the aquo ions in $HClO_4$ and HCl media was studied by Anderson and Bonner.[124] In perchlorate media the exchange obeyed the rate law

$$R = \left\{k_0 + \frac{k_h K_h}{[\text{H}^+]}\right\}[\text{Cr(II)}][\text{Cr(III)}]$$

where, at 25°C, $k_0 \leqslant 2.5 \times 10^{-5}$ l./mole sec. This rate law, like that for the ferrous-ferric exchange, is consistent with the mechanism

$$\text{Cr}^{+2}(\text{aq}) + {}^*\text{Cr}^{+3}(\text{aq}) \xrightarrow{k_0} \text{Cr}^{+3}(\text{aq}) + {}^*\text{Cr}^{+2}(\text{aq})$$

$$\text{Cr}^{+2}(\text{aq}) + {}^*\text{CrOH}^{+2} \xrightarrow{k_h} \text{CrOH}^{+2} + {}^*\text{Cr}^{+2}(\text{aq})$$

Since k_0 is small compared to $k_h K_h$, it will be assumed that the observed rate constant k is given by $k = k_h K_h$. The measured enthalpy and entropy of activation for k are 22 ± 2 kcal./mole and -4 cal./deg. mole, respectively. Using values of thermodynamic quantities listed in Latimer,[136] it is readily calculated that $K_h = 1.5 \times 10^{-4}$ M at 25°C and zero ionic strength and that $\Delta H_h^\circ = 16$ kcal./mole and $\Delta S_h^\circ = 40$ cal./deg. mole. Combining these values with the values pertaining to k, it is readily found that the enthalpy and entropy of activation for k_h are approximately 5 kcal./mole and -44 cal./deg. mole, respectively. These values are similar to those for the $\text{Cr}^{+2} + \text{Co(NH}_3)_5(\text{H}_2\text{O})^{+3}$ reaction. Since the Cr exchange reactions are not spin-forbidden and since they probably involve bridged inner-sphere mechanisms, the very negative value for $\Delta S_h^\ddagger$ is questionable.

The value of K_h in the previous paragraph was for a different ionic strength than the one at which k was measured. If the expression for the variation of the acid dissociation constant of $\text{Fe(H}_2\text{O})_6^{+3}$ with ionic strength is used for $\text{Cr(H}_2\text{O})_6^{+3}$ (since the charges are equal and the radii are so similar), then it is found that for unit ionic strength K_h is equal to 3.4×10^{-6} and 3.7×10^{-5} M at 0° and 25°C, respectively, and that $\Delta H_h^\circ = 15.6$ kcal./mole. The correction is negligible.

Effect of Cl$^-$

Free Cl$^-$ added to the reaction had negligible effect on the rate of Cr isotope exchange.[124] Therefore no activated complexes with Cl$^-$ in outer coordination shells, e.g., $[(\text{H}_2\text{O})_5\text{Cr(OH}_2) \cdots \text{Cl}^- \cdots (\text{H}_2\text{O})\text{Cr(H}_2\text{O})_5]^{+4}$ or a similar one with OH$^-$ ion in place of a water molecule, contributed measureably to isotope exchange.

Catalysis of exchange by Cl$^-$ in the first coordination shells of either ion should not occur to any appreciable extent because of the inert nature of $\text{Cr(H}_2\text{O})_6^{+3}$ toward Cl$^-$ substitution. For example, if the forward reaction

$${}^*\text{Cr(H}_2\text{O})_6^{+2} + \text{Cr(H}_2\text{O})_5\text{Cl}^{+2} \rightleftharpoons$$

$$\left\{\left[(\text{H}_2\text{O})_5\ {}^*\text{Cr}\overset{\displaystyle \text{H}}{\underset{\displaystyle \text{H}}{\overset{\displaystyle |}{\underset{\displaystyle |}{-\text{O}-}}}}\text{Cr(H}_2\text{O})_4\text{Cl}\right]^{+4}\right\}^{\ddagger} \rightarrow$$

$${}^*\text{Cr(H}_2\text{O})_6^{+3} + \text{Cr(H}_2\text{O})_5\text{Cl}^+ \quad (3\text{-}77)$$

contributes negligibly because of the slow formation of $Cr(H_2O)_5Cl^{+2}$, then the principle of detailed balance states that the reverse reaction will also contribute negligibly.

It is interesting to note that the forward reaction in Eq. 3–77 provides a mechanism for Cr^{+2}-catalyzed aquation of $Cr(H_2O)_5Cl^{+2}$ because the Cr(II) complex is substitution-labile. Since aquation of $CrCl^{+2}$ is very slow even in the presence of $Cr^{+2}(aq)$,[137] it must be concluded that chloropentaaquo complexes do not contribute to isotope exchange via H_2O- or OH^--bridged inner-sphere activated complexes or outer-sphere activated complexes.

Although electron transfer by H_2O- or OH^--bridged inner-sphere mechanisms is slow, electron transfer by a Cl^--bridged inner-sphere mechanism is relatively rapid. Taube and King[137] reported a value of 8.3 ± 1.7 l./mole sec. for the second-order rate constant at 0°C for isotope exchange between $Cr^{+2}(aq)$ and $CrCl^{+2}$. Isotope exchange occurred without net loss of Cl^- to the solution. Exchange was postulated to occur by

$$*Cr^{+2} + CrCl^{+2} \rightleftharpoons \{*Cr \cdots Cl^- \cdots Cr\}^{\ddagger} \rightarrow *CrCl^{+2} + Cr^{+2}$$

The rate of exchange of labelled Cl^- between $CrCl^{+2}$ and free Cl^- was greatly accelerated by the presence of $Cr^{+2}(aq)$.[137] The mechanism postulated was

$$*Cr^{+2} + Cl^- + CrCl^{+2} \rightleftharpoons \{Cl^- — *Cr \cdots Cl^- \cdots Cr\}^{\ddagger} \rightleftharpoons *CrCl_2^+ + Cr^{+2}$$

The third-order rate constant, k_3, for this reaction at 0°C was 0.5 l.²/mole² sec.

An upper limit for the formation constant of $CrCl^+$ can be obtained from the foregoing result. If the reasonable assumption is made that the catalysis of Cl^- exchange occurs through formation of $CrCl^+$ followed by a rate-determining reaction between $CrCl^+$ and $CrCl^{+2}$, then the third-order rate constant is given by

$$k_3 = k_2 K = 0.5 \text{ l}^2 /\text{mole}^2 \text{ sec.}$$

where k_2 is the second-order rate constant for the rate-determining step and K is the formation constant of $CrCl^+$. The value of the rate constant for the $Cr^{+2}(aq) + CrCl^{+2}$ reaction at 0°C, $\sim$9 l./mole sec., can probably be taken as a lower limit for k_2; this assumes that the free energy of $\{Cl—Cr \cdots Cl \cdots Cr\}^{\ddagger}$ is not greater than that of $\{Cr \cdots Cl \cdots Cr\}^{\ddagger}$ measured relative to $Cr^{+2}(aq) + Cl^- + CrCl^{+2}$, which is reasonable. Hence the upper limit for K is calculated to be

$$K = \frac{k_3}{k_2} = 5.5 \times 10^{-2} \text{ l./mole}$$

K is probably less than this, because k_2 is probably greater than 9 l./mole sec.

Effects of Anions Other than Cl^-

Ball and King[138] measured the rate of other electron-transfer reactions between $Cr^{+2}(aq)$ and CrX^{+2} in which X was F^-, Cl^-, Br^-, NCS^-, and N_3^-.

The X group was transferred. At $27°C$, $k(F^-) = 2.6 \times 10^{-2}$ l./mole sec., with $\Delta H^{\ddagger} = 13.7$ kcal./mole and $\Delta S^{\ddagger} = -20$ cal./deg. mole. At $0°C$, $k(Cl^-) = 9.1$ l./mole sec. in agreement with the value of Taube and King,[137] $k(Br^-) > 60$ l./mole sec. and, at $27°C$, $k(NCS^-) = 1.8 \times 10^{-4}$ l./mole sec. Recent measurements[139] for N_3^- gave $\Delta H^{\ddagger} = 9.6$ kcal./mole and $\Delta S^{\ddagger} = -23$ cal./deg. mole. The reactions with F^-, Cl^-, and N_3^- were pH-independent over the small ranges investigated; in other reactions pH dependence was not determined. As in the reactions of aquo chromous ion with penta-amminecobalt(III) ions, the rate constant shows a strong dependence on the nature of the bridging ion.

Chia and King[140] could not find evidence for a double F^- bridge in the inner-sphere activated complex of the $Cr^{+2}(aq) + cis\text{-}CrF_2^+$ reaction, but Snellgrove and King[141] found good evidence for such a transition state with $Cr^{+2}(aq)$ and $cis\text{-}Cr(N_3)_2^+$.

Catalysis of the formation of Cr(III)—X complex ions by $Cr^{+2}(aq)$ has been studied by Hunt and Earley.[142] The observed order of reaction rates was the same as the order of stability of the resulting Cr(III) complexes, namely, EDTA > pyrophosphate > citrate $\simeq$ phosphate > F^- > tartrate > NCS^- > SO_4^{-2}. Complex formation between Cr^{+2} and X, followed by electron transfer between $Cr^{+3}(aq)$ and the Cr(II)—X complex in an H_2O-bridged activated complex similar to that given in Eq. 3–77, was postulated.[142]

The $Cr^{+2}(aq) + Cr(NH_3)_5X^{+2}$ Reactions

These reactions are analogous to the reactions with pentaammine-cobalt(III). The net reaction was[81]

$$Cr^{+2}(aq) + Cr(NH_3)_5X^{+2} + 5 H^+ \rightarrow CrX^{+2} + Cr^{+2}(aq) + 5 NH_4^+$$

The rate of disappearance of the pentaamminechromium(III) was given by

$$R = k_1[Cr(NH_3)_5X^{+2}] + k_2[Cr^{+2}][Cr(NH_3)_5X^{+2}]$$

for $X = F^-$, Cl^-, Br^-. The enthalpies of activation for k_2 in this same order were 13.4, 11.1, and 8.5 kcal./mole, and the entropies of activation were -30, -23, and -33 cal./deg. mole. The enthalpies of activation are similar in magnitude to those for the $Fe^{+2} + Fe(III)$ exchange catalyzed by the same ions and to those for F^- and N_3^- bridges in the $Cr^{+2} + CrX^{+2}$ exchange. The entropies of activation for F^- and Br^- given above are somewhat more negative than those for the F^-- and N_3^--bridged $Cr^{+2} + CrX^{+2}$ exchanges, but the difference may be within the experimental error. These entropy changes should be compared with those for the $Fe^{+2} + FeN_3^{+2}$ reaction discussed earlier. It will be recalled that the kinetic evidence is consistent with the interpretation that at low temperatures formation of the dinuclear complex, $\Delta H^{\ddagger} = 13.9$ kcal./mole, $\Delta S^{\ddagger} = 7.0$ cal./deg. mole, was rate-determining whereas at higher temperatures electron transfer in the dimer,

$\Delta H^{\ddagger} \sim 5$ kcal./mole, $\Delta S^{\ddagger} \sim -24$ cal./deg. mole was rate-determining.[25] Thus entropies of activation of approximately -20 to -30 cal./deg. mole seem to be associated with reactions between complex ions having small ligands which are known to proceed by an inner-sphere mechanism and which are multiplicity-allowed. However, this is not a very cogent reason for postulating inner-sphere mechanisms for the iron isotope exchanges catalyzed by F^-, Cl^-, Br^- (see Table 3–1), since similar entropies of activation can be expected on the basis of outer-sphere mechanisms for Fe exchanges with charge products of $+4$ and $+6$.[18] The entropies of activation for inner-sphere activated complexes can be considered to be due chiefly to an entropy change associated with dimer formation (from loss of motional entropy and from the coulombic repulsion term) and an entropy change associated with rearrangement of outer solvation shells. The entropy change associated with inner coordination shells is probably negligible. First-coordination-shell ligand–metal ion bonds are strong, often very strong, and the system of reactant ions would change entropy very little when ligand-metal distances are changed 5–10 percent or when bond lengths within a ligand are changed by this amount. (See the calculation for the aquo ferrous ion $+$ tris-1,10-phenanthroline Fe(III) reaction. The change of Fe(II)—H_2O bond length was less than 4 percent.)

Effect of D_2O

The rate of reaction of the aquo chromous ion $+$ chloropentaammine Cr(III) reaction was determined[81] in 86 percent D_2O at $26°C$. The ratio k_H/k_D was found to be 1.3, the same as the value found for the aquo chromous ion $+$ hexaammine Co(III) reaction at $37°C$. Since a Cl^- bridge is formed, the D_2O effect presumably arises mainly in the first coordination sphere of Cr^{+2}, as in the reaction with the Co(III) complex.

4

Energy Surfaces

FORMAL ASPECTS OF KINETICS

It will be assumed throughout that the theory of absolute reaction rates[143] is applicable and that the rate constant may be written as

$$k = \kappa \frac{RT}{Nh} \exp\left(\frac{-\Delta F^{\ddagger}}{RT}\right) = \kappa \frac{RT}{Nh} \exp\left(\frac{\Delta S^{\ddagger}}{R}\right) \exp\left(\frac{-\Delta H^{\ddagger}}{RT}\right) \quad (4\text{--}1)$$

The free energy, enthalpy, and entropy of activation are, by definition, based on partition functions from which the factor for the degree of freedom corresponding to motion in the reaction coordinate of the activated complex has been removed. The transmission coefficient, κ, is much less than unity if the reaction is non-adiabatic and of the order of unity if the reaction is adiabatic. The other symbols have their usual meaning.

It is customary to represent chemical reactions by the movement of a point on potential-energy surfaces, a full display of which requires as many coordinates as there are vibrations plus one. Embarrassment is avoided by assuming that, of all the pathways which lie between the portions of the surfaces representing reactants and products, one pathway is much more important than any other usually because it has no point at which the potential energy is as high as it is at some point on each of the other pathways. The single important path is projected onto two dimensions to show variation of potential energy with position of the reactants along the reaction coordinate.

The potential energy at each point of a potential-energy surface is a sum of electronic kinetic and potential energies and of nuclear potential energy. Ideally, the potential energy at each point is the eigenvalue of the fixed-nuclei Hamiltonian operator defined by

$$\mathsf{H}_n = -\frac{1}{2}\sum_i \nabla_i^2 + \sum_{i>j} \frac{1}{r_{ij}} + \sum_{k>l} \frac{Z_k Z_l}{r_{kl}} - \sum_{i,k} \frac{Z_k}{r_{ik}} + \text{M.T.} \quad (4\text{--}2)$$

at the nuclear configuration represented by the point n, for the state Ψ_n. If the state Ψ_n is the electronic ground state for a given nuclear configuration

n, the eigenvalue E_n° will be the lowest eigenvalue of H_n and the point in question will lie on the lowest potential-energy surface; if the state Ψ_n is an excited electronic state for the given nuclear configuration n, the eigenvalue E_n will be greater than E_n° and the point in question will lie on an upper potential-energy surface. As the nuclear configuration is varied from point to point, the potential-energy surfaces for the ground and excited states can be traced out with the aid of the adiabatic principle of Ehrenfest,[144] which states that a system will always remain in an eigenstate if the surroundings are changed sufficiently slowly.

In Eq. 4–2 atomic units are used, i and j refer to electrons only, k and l refer to nuclei only, Z_k is the charge on nucleus k, and M.T. is an abbreviation for "more terms"! In writing Eq. 4–2 for H_n the Born-Oppenheimer approximation is used; this approximation assumes that nuclear motions in ordinary molecular vibrations are so slow compared to electronic motions that the electrons are not affected by nuclear momenta, only by nuclear positions.

ZERO-ORDER AND FIRST-ORDER SURFACES

For many-electron systems, the exact eigenfunctions and eigenvalues of H_n are unknown. Consequently, the exact potential-energy surfaces on which the nuclei are to move are not obtainable. Approximate potential-energy surfaces are obtained by using approximate wave functions. When a zero-order wave function, say the electronic wave function of the separated ground state reactants Ψ_R, is used to calculate energies as in

$$E_R = \int \Psi_R H_n \Psi_R \, dV = \langle \Psi_R | H_n | \Psi_R \rangle \tag{4–3}$$

for all nuclear configurations of interest, a zero-order energy surface, denoted by E_R, is obtained. Another zero-order energy surface, denoted by E_P, is obtained when the electronic wave function of the separated products is used. For certain nuclear configurations E_R will be below E_P, i.e., the reactant electron distribution will have a smaller energy than the product electron distribution for these nuclear configurations. For other nuclear configurations, E_P will be below E_R. Where the zero-order energy surfaces intersect there is a set of nuclear configurations for which $E_R = E_P$. Over a portion of the intersection region where the reactants are not too far apart, the two zero-order states, Ψ_R and Ψ_P, can interact. It is assumed that the interaction is not rigorously spin-forbidden.* The interaction arises in the same

* If the interaction between the two zero-order states is rigorously spin-forbidden, the reaction will not occur. The Wigner-Wittmer correlation rules allow one to determine whether the interaction is spin-forbidden or not if the spins of the reactants and products are known. When the interaction is spin-forbidden, sufficiently large spin-orbit coupling terms in H_n can still permit interaction between the zero-order states, since the total spin of reactants or products is not a rigorous quantum number in this case.

way that "resonance" usually arises, namely, that two or more electron distributions have similar energies for a given nuclear configuration. Even if Ψ_R and Ψ_P were exact electronic wave functions for the separated reactants and products, respectively, they individually no longer serve to describe the system. To describe the system an approximate wave function, which is a linear combination of Ψ_R and Ψ_P as in

$$\Psi = a\Psi_R + b\Psi_P \tag{4-4}$$

is usually used. The Rayleigh-Ritz variation method can be applied to this linear combination to give the improved first-order energies, E_+ and E_-, of the two new states, Ψ_+ and Ψ_-, which result from the interaction of the two zero-order states. Where $E_R = E_P$, the expressions for the first-order energies are

$$E_\pm = \frac{E_R \pm H_{RP}}{1 \pm S} \tag{4-5}$$

in which S, the overlap integral, is equal to $\langle \Psi_R | \Psi_P \rangle$ and the interaction energy H_{RP} is equal to $\langle \Psi_R | H_n | \Psi_P \rangle$. At points close to the intersection, the expressions for $E_\pm$ are somewhat more complicated and are obtained by solving the 2×2 secular determinant when E_R is not equal to E_P. The interaction energy $|H_{RP}|$ decreases rather quickly with increasing distance from the intersection so that the first-order energy surfaces E_+ and E_- quickly approach the zero-order surfaces E_R and E_P in regions not too distant from the intersection region.

For all nuclear configurations represented by the reactant and product surfaces $E_+ \leqslant E_R$ or E_P, whichever is lower, and $E_- \geqslant E_R$ or E_P, whichever is the higher. The upper portions of the E_R and E_P surfaces form the first-order surface E_- and the lower portions of the E_R and E_P surfaces form the first-order surface E_+. The first-order surfaces do not cross. Wherever the zero-order states Ψ_R and Ψ_P are allowed to interact appreciably, the energy difference between the first-order surfaces is $E_- - E_+$; in the intersection region of E_R and E_P this difference is given by

$$\Delta E = E_- - E_+ = \frac{2E_R S - 2H_{RP}}{1 - S^2}$$

or, if $S \cong 0$,

$$\Delta E = -2H_{RP} \tag{4-6}$$

THE TRANSMISSION COEFFICIENT

When the interaction energy $|H_{RP}|$ is greater than a few tenths of a kilocalorie per mole the splitting, ΔE, of the first-order surfaces will be sufficiently large so that the point, representing the nuclear configuration and electronic state, will remain on the lower first-order energy surface as it moves along the reaction coordinate. The change from reactants to products is adiabatic in the Ehrenfest sense. The transmission coefficient, κ, in Eq. 4–1 is of the order of unity.

When the interaction energy $|H_{RP}|$ is less than a few calories per mole, the splitting, ΔE, of the first-order surfaces is so small that as the point passes through the configuration of an activated complex the electronic wave function will change from $\Psi_+ \cong \Psi_R$ to $\Psi_- \cong \Psi'_R$, i.e., the point will follow the zero-order surface E_R through the activated complex region. Most activated complexes will then not lead to product and if there is a solvent cage to confine the reactants, the point may pass through the very small interaction region many times before it succeeds in making the transition. The change from reactants to products is non-adiabatic and $\kappa \ll 1$.

FREE-ENERGY SURFACES

Energies calculated with the Hamiltonian operator in Eq. 4–2 and approximate wave functions for the complex reactants are unreliable at present. On the other hand, free-energy changes calculated from electrostatic theory and force constants would seem to be in much better agreement with experimental values although still far from satisfactory. Entropy changes are included in the free-energy changes, and the calculation of these is less reliable partly because of the unknown variation of dielectric constant with temperature in the neighborhood of ions. Enthalpy changes obtained by combining the calculated free energies and entropies reflect uncertainties in the latter two quantities. Hence it seems preferable to plot free-energy curves rather than enthalpy curves. In the following paragraphs, a more detailed account of free-energy surfaces will be given than was given for the energy surfaces in the preceding paragraphs.

For a concrete example, we will assume that the hexaaquo ferrous and ferric isotope exchange reaction proceeds by an outer-sphere mechanism. For simplicity, it will be assumed that the nuclear configuration of the activated complex is achieved through simple spherically symmetric breathing vibrations of the ligands following an initial positioning of the two central metal ions relative to each other. Rearrangements of the second and outer hydration shells will accompany these vibrations and these will be considered as linked to the inner-shell rearrangements, so that the positions of all the water molecules can be described in terms of two coordinates, one for each ion. Thus very many of the configurations leading to transition states which do not have spherically symmetric hydration around the two ions have been eliminated. To further simplify, we consider only those in the group of remaining configurations through which the system may move to the transition state through equal but opposite changes in the two distance parameters. The activated complex thus produced will have identical, symmetrical hydration around the two ions. The path through these configurations can be described in terms of a single coordinate ρ, which simultaneously measures expansion of the $Fe^{+3}(aq)$ coordination shells and compression of those of $Fe^{+2}(aq)$, these motions being so phased as to reach an activated complex of symmetrical configuration. Such complexes are not the only ones which are likely to be important, but they are the easiest

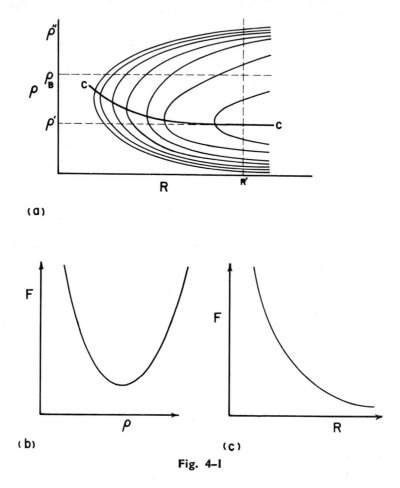

(a)

(b)　　(c)

Fig. 4-1

to employ in the present exposition. For these symmetric activated complexes to make an important contribution to the outer-sphere mechanism, the distance of approach R, of the two ions must be sufficiently small so that, the interaction energy H_{RP} does not vanish.

Curves showing the variation of free energy with ρ and R are shown in Fig. 4-1. Figure 4-1b is a typical cross section of the surface through a line of constant R. At the minimum on this curve, the hydration shells are in their equilibrium configurations around the ions for the given value of R. The line of these configurations for all R is CC in Fig. 4–1a. Figure 4–1c shows a curve roughly describing the variation of free energy with R for fixed ρ. In interpreting Fig. 4–1a, let us assume that moving downward on the ρ coordinate corresponds to a "squeezing-in" of the hydration shells around Fe^{+3} and a simultaneous expansion of the hydration shells around Fe^{+2}, whereas moving upward on the ρ coordinate corresponds to a "squeezing-in" of the hydration shells around Fe^{+2} and a simultaneous expansion of the hydration shells around Fe^{+3}. At large R, the equilibrium values of ρ are on the lower part of the ρ axis to indicate that all Fe^{+3}—H_2O distances are less than the corresponding Fe^{+2}—H_2O distances. If an electron was transferred from Fe^{+2} to Fe^{+3} when $R = R'$ and $\rho = \rho'$, the resulting Fe^{+2} would have a "squeezed-in" set of hydration shells and the resulting Fe^{+3} would have a set of expanded hydration shells corresponding to $\rho = \rho''$. The latter nuclear arrangement has a much higher free energy than the former, and transfer could not be made without an input of energy during the time required for the electron transition. Somewhere in between ρ' and ρ'', say ρ_B, all Fe^{+3}—OH_2 distances equal the corresponding Fe^{+2}—OH_2 distances. The activated complexes we are discussing lie along the line $\rho = \rho_B$. For each value of ρ_B we can draw a curve similar to that of Fig. 4–1c. It must be remembered that all paths of configuration change leading to the *same* activated complex with configuration ρ_B and R by routes in which motions of the water molecules are not synchronized have been ignored. This simplification is convenient for picturing the true, complex process which occurs. However, the calculation of the rate constant for this activated complex defined by ρ_B and R depends only on the description of the activated complex and not on the pathways by which it was reached. Thus, the fact that the pathways to a given activated complex have been greatly restricted here does not mean that the calculation of the free energy of activation for that activated complex need be complicated by separate consideration of all possible paths to that activated complex.

In Fig. 4–2a, a number of curves similar to the one in Fig. 4–1b are given at various fixed values of R for the reactant electron distribution, $Fe^{+2}(aq) +$ *$Fe^{+3}(aq)$; in Fig. 4–2b, the same curves are given for the product electron distribution, $Fe^{+3}(aq) +$ *$Fe^{+2}(aq)$. In both figures, ρ_B is the value of ρ for identical hydration configurations around both ions. In both figures, the zero-order free-energy surfaces are under consideration. These surfaces must be added to form a single figure. This has been done in Fig. 4–3 which shows one F–ρ cross section at a fixed value of R.

At large values of R, the cross sections of the first-order surfaces are also those of the zero-order surfaces (as shown in Fig. 4–4a) because the interaction energy $|H_{RP}|$ falls off rather rapidly with increasing R. At smaller values of R, the interaction between the zero-order states increases; in the resulting

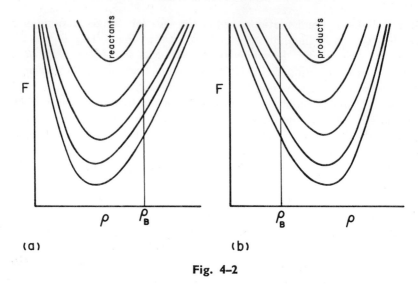

Fig. 4–2

pair of first-order surfaces, one is everywhere of higher free energy (surface *2* in Figs. 4–4b and c) than the other (surface *1* in the same figures).

Figure 4–5 presents a view looking down on the total figure in which only the lower parts of the two zero-order surfaces are shown. The solid straight line in the center of this figure is the locus of configurations in which the hydration about the two ions is the same but for which $|H_{RP}|$ is vanishingly small as a result of R being too large. At smaller R, the reactant and product states do interact; the dotted line is the locus of configurations of acceptable activated complexes to which our attention is directed. Acceptable reaction paths can cross at any point along the dotted line but not across the solid line. What might be a typical reaction path is shown by the preambulating line of this figure; the point marked $X_i^{\ddagger}$ is the transition state of this

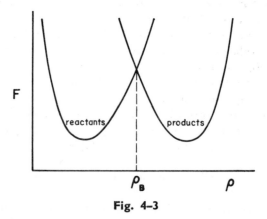

Fig. 4–3

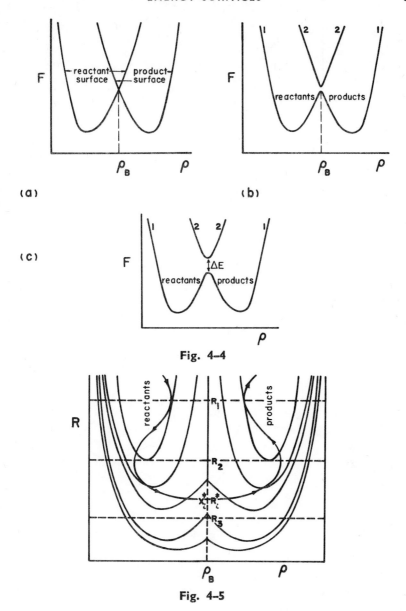

(a)

(b)

(c)

Fig. 4–4

Fig. 4–5

particular path. Other paths will cross the dotted line at other points. The observed rate constant will contain contributions from all such paths.

In Fig. 4–6, the F–R plane at $\rho = \rho_B$ is shown. The solid line of Fig. 4–5 extends from the extreme right of Fig. 4–6 to the point where surfaces 1 and 2 separate at $Y^{\ddagger}$; $Y^{\ddagger}$ is the transition state with the largest allowed value of

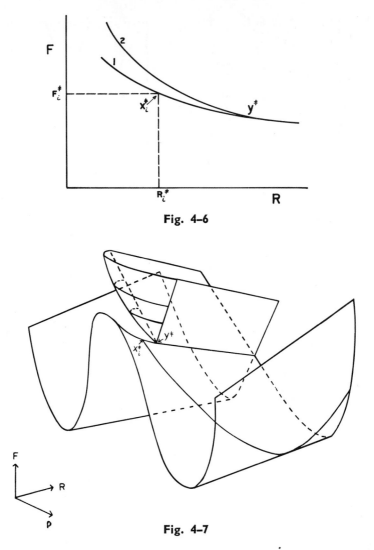

Fig. 4–6

Fig. 4–7

$R^{\ddagger}$.   From this point to the left, surface 1 always lies below surface 2.   In the present reaction, the complex $X_i^{\ddagger}$ may be a more important one than $Y^{\ddagger}$ because of the very small $|H_{RP}|$ for $Y^{\ddagger}$, although the free energy of activation for $X_i^{\ddagger}$ is larger than that for $Y^{\ddagger}$.

In Fig. 4–7, an attempt has been made to present a perspective view of the first-order surfaces, showing both where they coincide with the zero-order surfaces and where the upper surface splits away from the lower surface. The wavy line in this figure is meant to show a portion of the preambulating line of Fig. 4–5.

5

Non-Adiabatic Electron Transfer

If, for some reason, the interaction matrix element H'_{RP} between the zero-order states for reactants and products is very small, the reacting system, as it passes through the configuration of the activated complex, seldom crosses from one zero-order surface to the other and reaction does not occur. Looked at in another way, we could say that the system fails to stay on the lower first-order surface (surface 1 of Fig. 5–1) and "jumps" across the

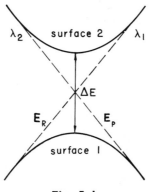

Fig. 5–1

small interaction energy gap to the higher first-order surface (surface 2 of Fig. 5–1). Either picture is appropriate, and both have been used to discuss the problem. Unfortunately, the term "non-adiabatic" has been used rather indiscriminately. In some cases it has been used to designate crossing between zero-order surfaces, and in other cases for crossing between first-order surfaces. To be consistent with our choice of definition for the term, we shall refer to crossing between zero-order surfaces as being the central characteristic of non-adiabatic reactions, since a poor crossing probability is equivalent to a poor transmission coefficient for electron transfer. Even this definite statement of the meaning of the term is not a good operational definition since there is no general way to separate the transmission coefficient from the entropy factor in the temperature-independent part of the rate constant and thus no general way to determine whether the transmission coefficient is large or small.   Indeed, it has not been possible thus far to establish that the transmission coefficient for any electron-transfer reaction is actually so small that the reaction must be said to be non-adiabatic. Certainly, most of the electron-transfer reactions of complex ions which have been measured are adiabatic.   Candidly, we can hope that they all turn out to fall in this category since there are probably few problems in quantum mechanics quite so difficult to treat quantitatively as the transmission coefficient for non-adiabatic reactions in liquid phases.   It may be suggested that there are no non-adiabatic reactions in solution, and this suggestion is not so unreasonable as it might sound when we consider that an interaction energy of a few hundred calories per mole or less is all that is required for an adiabatic reaction as we have defined the term.

It has often been suggested that the range of rate constants for electron transfer is so large that the slowest reactions must be non-adiabatic.   The rate constants for electron-transfer reactions thus far measured cover a range of sixteen orders of magnitude.   However, this range of rate constants is equivalent to a difference of only 22 kcal. between maximum and minimum free energies of activation.   This is less than 1 electron volt and is not so large that one feels forced to suggest that the slowest processes are non-adiabatic;   rates of adiabatic reactions readily cover this range.   However, the unexpected slowness of a few reactions does suggest non-adiabatic possibilities.   For example, the reaction

$$Cr^{+2}(aq) + Co(NH_3)_6^{+3} \rightarrow Cr^{+3}(aq) + Co(NH_3)_6^{+2} \qquad (5\text{--}1)$$

which has a rate constant of 9×10^{-5} l./mole sec., might be expected to be faster when compared with other reactions of similar complex ions.   There is a large change in multiplicity in this particular reaction that is suggestive of a spin-conservation problem which could make electron transfer rate-limiting.   The same thing is true for the oxidation of hemoglobin by oxygen, which is a very slow reaction.   In inner-sphere mechanisms in which the reactants are bridged by one or more long ligands, it is possible that the "electrical conductance" through the bridge is poor and these reactions may also be non-adiabatic.   However, in none of these cases is it yet established that electron transfer within the activated complex is rate-limiting.   Fortunately, in some reactions it should be possible to test this possibility by experiments.   Particularly in homonuclear reactions for which the electrostatic factor in the entropy of activation is sufficiently small to allow good estimation of the electrostatic contribution to the activation entropy, a non-adiabatic process should be revealed by an anomalously large and negative value of the apparent entropy of activation since the transmission coefficient appears in this term. Of course, it is also necessary to know the mechanism of the reaction. There is reason to suppose that some oxidation-reduction reactions of complex ions depend on atom transfer rather than direct electron migration. This is consistent with knowledge of organic reactions in which hydrogen-atom and hydride-ion transfer is known.

Since there are, in fact, a few rather large and negative activation entropies in complex-ion, electron-transfer reactions thought to involve direct electron migration, we cannot dismiss the non-adiabatic category out of hand; it is necessary to consider the theories and peculiarities of such processes though we shall only be able to do so in a formal and preliminary way. The basic problems in theoretical treatments lie not so much in the formal theory itself but in applications. It is first necessary to have a complete description of the elementary reactions. Then the important perturbation must be determined without, if possible, the use of theory; and finally, one must be able to write wave functions for reactants and products as tractable mathematical expressions. In particular, the latter requirement is quite beyond quantum-mechanical methods at present. Only the most elaborate theoretical treatment of the transmission coefficient can be expected to provide results reliable to better than several orders of magnitude. Simpler calculations may be useful insofar as they give estimates which are near unity or very small, but they will not allow a quantitative separation of apparent activation entropy into the transmission-coefficient contribution and the true entropy of activated-complex formation. It is probable that the transmission-coefficient problem in electron-transfer reactions should not be approached through reactions in solution. From a theoretical point of view, several simpler situations can be suggested. For example, Rice[145] has suggested the possibility of bombarding weak molecular crystals of relatively simple organic or inorganic substances in their oxidized state with electrons, followed by a study of electron migration in these crystals. The fixed, known geometry, plus the possibility of writing suitably accurate wave functions, makes this kind of experiment much more attractive from the theoretical point of view.

A brief historical survey of the treatment of non-adiabatic reactions will be useful and will provide an introduction to methods and concepts necessary for the following chapter. By far the most extensive literature on the subject is that pertaining to scattering experiments in gas phases. For the most part, this literature is not directly relevant at the level of the present discussion, though we shall have to refer to some theoretical conclusions presently available only for simple scattering problems in gases.

Historically, there are two general approaches ultimately applicable to the problem of calculation of transmission coefficients for electron-transfer reactions in the liquid phase. The older, which is more appropriate in light of modern developments in theory and more powerful, is that of time-dependent perturbation theory. Libby[1] used a very simple treatment of this sort in 1952, but the well-known transition probability expression usually attributed to Landau[146] and Zener[147] is much older and still formally applicable to the transmission-coefficient problem in non-adiabatic electron-transfer reactions.

The second approach, introduced for this problem by R. J. Marcus, Zwolinski, and Eyring,[2] is to calculate the transmission coefficient for penetration of the potential-energy barrier restricting electron transfer between reactants by a wave packet estimated for the migrating electron. A single plane wave has been used. Such an approximation is very crude and there is no reason to believe that a more refined treatment involving educated guesses as to the composition of the wave packet would much improve matters. Hence this approach has largely been replaced by that based on time-dependent perturbation theories.

ELECTRON TUNNELLING TRANSMISSION COEFFICIENTS

It is a well-known result of quantum mechanics that particles may pass through potential-energy barriers of finite height and width where their total energy is less than their potential energy. Such passages are forbidden in classical mechanics. Marcus, Zwolinski, and Eyring[2] assumed that an electron passed through a potential-energy barrier when it passed through the region of space occupied by water molecules of hydration separating two hydrated metal ions. Orbitals on each metal ion represented allowed regions of space between which the electron must tunnel in order for the system to pass from the reactants' to the products' surface at the nuclear configuration of the activated complex. The probability of the system passing from the reactants' to the products' surface, i.e., κ, was taken to be equal to the tunnelling transmission coefficient of the electron through the potential-energy barrier between the orbitals. For ions of like charge the free energy of activation, $\Delta F^{\ddagger}$, and κ both decrease with increasing separation between the metal ions but with opposite effects on the rate constant given in Eq. 4–1. The distance between the metal ions in the activated complex was determined by maximizing the rate constant with respect to this distance. The potential barrier was estimated from point-charge coulombic interactions. In this treatment as in other tunnelling treatments, the effective rate constant for electron transfer in the activated complex should be the product of κ, the tunnelling transmission coefficient, and the frequency with which the electron strikes the barrier. This fact has been ignored in these theories, so that the calculated rate constants thus far reported may be expected to be highly unreliable. Furthermore, the details of calculating realistic values of the free energy of formation of the activated complex have been omitted. The method of Marcus and Hush discussed in Chapter 6 should thus be used to complement the actual calculation of κ in tunnelling theories.

Laidler[5] and Sacher and Laidler[16] modified the original tunnelling treatment to include a variation of dielectric constant with field strength, a calculation of the reorganization energy of the inner and outer coordination shells of the reactant metal ions, use of a rectangular as well as a triangular potential-energy barrier, and a weighted average of the apparent free energies of activation over a range of metal ion–metal ion distances. The apparent free energy of activation is obtained from Eq. 4–1 by setting the transmission coefficient equal to unity. The results obtained[16] for the $Fe^{+2}(aq) + Fe^{+3}(aq)$ reaction were that the metal ion–metal ion distance was approximately 4.2 Å and that the weighted apparent $\Delta F^{\ddagger}$ values for the rectangular and triangular potential-energy barriers were 15.9 and 15.2 kcal./mole, respectively, compared to the experimental value of 16.8 kcal./mole. The distance of 4.2 Å corresponds closely to only one water molecule separating the metal ions, as would be the case if the reaction proceeded by a water-bridged inner-sphere activated complex. However, the method of calculating the reorganization energy was not consistent with this result since the energy for replacing an inner-shell water ligand in one ion with a water ligand essentially shared by the inner shells of both ions was not included.

The tunnelling theories to date have been applied only to systems where the potential energy of the electron was the same on each side of the potential-energy barrier, for example, homonuclear electron-transfer reactions with correctly oriented inner coordination and outer solvation shells. They have not been applied to heteronuclear electron-transfer reactions where the potential energy of the electron is different on each side of the barrier. The height of the energy barrier and the total energy of the electron are difficult quantities to estimate reliably. Even if the potential energy and total energy of the electron in the relatively simple, free gaseous metal ion were known, these energies are greatly affected by the ligand field of the co-ordination spheres. The ligand field gives rise to a large spherically symmetrical and smaller directed components of the potential energy depending on the symmetry of the complex ion. The effects of ligands on Condon-Shortley parameters, on effective charge of the metal nucleus, and on the size or extent of an orbital cannot be predicted with accuracy. Success may occasionally crown efforts to calculate a total energy but such calculations are not to be relied upon and are not generally useful.

The formulation of the tunnelling probability and its dependence on potential energy, total energy of an electron and on barrier height and width does not permit one to take into account the effect on κ of a partially spin-forbidden or symmetry-forbidden transition. Here interactions such as spin-orbit coupling and vibronic coupling must be considered, and these interactions are not easily included in the tunnelling formalism. However, it may be possible to use tunnelling expressions of simple form to estimate a rough order of magnitude of H'_{RP} as a preliminary test for adiabaticity.

THE PERTURBATION APPROACH

Nature of the Perturbation

The abundant literature on scattering problems of physics is based on some form of perturbation theory, usually time-dependent. The methods include Dirac's method of variation of constants, the distorted wave, and the method of perturbed stationary states, all of which appear to be applicable to electron-transfer (rearrangement) collisions in condensed phases. In physical scattering problems, almost invariably the potential energy or kinetic energy associated with the coordinate of relative motion of the incoming particle and the scattering center is assumed to be the perturbation mixing zero-order states to effect rearrangement. In solution reactions the effective perturbation is not so easily distinguished. It has been conventional to divide perturbations between the categories "strong" and "weak," but it is less confusing here to call them "slow" and "fast." Slow perturbations cannot produce a net change in quantum state since the system perturbed is always in nuclear and electronic equilibrium with the perturbation. There is further confusion possible at this point, in that such processes are adiabatic in the Ehrenfest sense and the perturbations thus quite reasonably are called "adiabatic perturbations"; let us simply stick to the term "slow." Perhaps a few additional words are necessary about such perturbations since in the usual considerations changes in at least nuclear coordinates are still slow with respect to a "slow" perturbation. In our reactions a characteristic slow perturbation is that supplied by the fields of the reactants as they move toward or away from each other ($\sim 10^{-9}$ sec.). The nuclear motions which establish the condition of degeneracy in the activated complex are, however, fast with respect to diffusional times. Hence the slow perturbation associated with diffusion can only be considered to be an additional time-independent term in the potential energy. Such perturbations thus appear to be of little interest to us in connection with the application of time-dependent perturbation theory. However, the mutual perturbations of the reactants mix the wave functions of the reactants and those of the products mix the wave functions of the products. It may then be possible to form zero-order wave functions for the collision complexes (cf. page 90) by time-independent perturbation theory, using as the starting functions those of either reactants or products at infinite separation (since these will be known with more accuracy than the wave functions of the collision complexes themselves), and the interreactant terms of the full Hamiltonian operator as the perturbation. In the remainder of this chapter we shall do just this, but it is important to note that despite the reasonableness of this procedure, very little is really known about the perturbation problem involved* and difficulties

---

* There is another complication which appears when this process is used to form expressions for the wave functions of the collision complexes. This is a consequence

may arise. For example, the fast perturbations are probably quite weak in most electron-transfer reactions and the slow perturbations strong or at least stronger. It is possible that the latter are in some cases "sufficiently fast" to be effective in mixing the zero-order states of the reactants' and products' activated-complex wave functions to provide the coupling for electron transfer. In other words, a large but slow perturbation may still change by a significant amount during the lifetime of the activated complex. However, this does seem unlikely for diffusional motions in condensed phases for which the problem is totally different from fast collision processes in gas phases. Parenthetically, it may be noted that it is not always ground states of reactants or products which need to be mixed for reaction to occur since if there is a large positive over-all free-energy change, the activated complex can be and often must be formed from reactants one or more of which has been previously excited to a higher electronic state. By similar reasoning, we expect one or more electronically excited products if the over-all free-energy change is a large negative quantity. Interesting examples of the latter phenomenon have been provided by Hercules[150] and by Chandross and Sonntag[151] (see p. 97) and must be a common occurrence for both electronic and vibrational states in gas-phase molecular beam processes.

Another conclusion from this discussion of perturbation times is that the times involved in the formation and destruction of the activated complex itself cannot be much different from those responsible for the mixing perturbation. It thus becomes very difficult to distinguish zero-order states of the activated complex from first-order states on any time scale, and it appears possible that the perturbation is indistinguishable from the solvent and ligand rearrangements which produce the activated complex. There is then

---

of the fact that the zero-order Hamiltonian to be used in the time-dependent treatment of the reactants' activated complex is different from the corresponding operator for the products' activated complex so that the zero-order wave functions are never strictly orthogonal. This is a well-known problem in rearrangement collision theory, called the "post-prior Hamiltonian problem." It has not been examined in detail for the type of reactions under consideration and might provide a means whereby slow perturbations can yet be effective in the mixing of these zero-order functions. We have carried through a reasonable, though quite approximate, investigation of the slow perturbation case and conclude that the usual limitations of slow perturbations apply. As regards fast perturbations, we conclude that the formal treatment (see Wu and Ohmura [148, pp. 192 et seq. and 293], and Massey [149]) of the post-prior problem is adequate for reactions of present interest. More confident analysis of the problem is probably required.

The post-prior problem will always appear in some form if the collision-complex wave functions are formed from those of the separated reactants, and this fact may ultimately eliminate this useful method for forming the zero-order wave expressions. The positions of the transferring electron on reactants and on products are, however, such as to introduce a close approach to orthogonality, so that the post-prior problem is unlikely to require consideration in theoretical treatments of condensed-phase electron-transfer reactions except at the most refined levels. At present, all practical treatments of the transmission coefficient contain the post-prior problem, often implicitly. It is present, for example, in the treatment of Levich and Dogonadze (pp. 145 et seq.; cf. p. 129).

no reason to assume that the perturbation is small, i.e., less than the upper limit set for non-adiabatic reactions, and we must conclude that it is doubtful that non-adiabatic electron-transfer reactions exist. If, however, the perturbation is after all a small one, its time cannot differ by much more than one power of ten from the lifetime of the activated complex and is probably of the same order. Then simple perturbation theory is not applicable, since the customary manner of treating this problem is to consider that the zero-order states are those of the activated complex produced in an adiabatic process. This procedure is, for example, that of Levich and Dogonadze discussed later in this chapter. A necessarily more realistic approach would be to consider the states of the reactants and products when brought to the internuclear distance of the activated complex, but before solvent and ligand reorganization, as the zero-order states. We shall call these states the *reactants'* and *products' collision complexes*, respectively. With this more realistic choice, the solvent and ligand rearrangements become the perturbation; but the energies associated with these rearrangements are frequently large, so that simple first-order perturbation theory is unlikely to be satisfactory. The problem then becomes very difficult, particularly since the transmission coefficient cannot be simply separated out as a factor in the rate constant but is intimately associated with the energy of solvent and ligand rearrangement, and the latter cannot be treated in the relatively simple fashion of Marcus and Hush to be discussed in Chapter 6. We are thus left in an embarrassing position, not only because our arguments have led us to believe that non-adiabatic electron-transfer reactions are improbable, but also because we are unable to accept the validity of conventional applications of perturbation theory to this problem. At the same time, the problem of the transmission coefficient in our reactions is so little understood that we cannot be totally certain that these preliminary arguments are themselves valid. In order to avoid future embarrassment, it seems necessary to adopt the following course of action: Let us remember the arguments just advanced but proceed to examine the prevailing point of view regardless of these arguments. This means we shall assume that perturbation theory is applicable, and we shall usually distinguish the adiabatic process of forming the activated complex by solvent and ligand rearrangements from the small perturbation which mixes these zero-order states. To develop these arguments, we shall have to consider four different sets of states, which must be clearly distinguished. They are:

1. The reactants or products at infinite separation from each other usually in equilibrium states. These are the infinite-separation states.
2. The reactants or products at internuclear coordinate values characteristic of the activated complex but with equilibrium solvent and ligand geometry. These are the collision-complex states.
3. The reactants' and products' activated-complex states degenerate in

energy and with full solvent and ligand rearrangement. These are the zero-order states of the activated complex.

4. The perturbed states of the reactants' and products' activated complexes mixed by the effective perturbation. These are the first-order states of the activated complex.

No matter what the approach, it is necessary to satisfy the Franck-Condon requirement in the first-order states of the activated complex. If the mixing perturbation is very small, as is necessary for non-adiabatic electron-transfer reactions, the zero-order states also satisfy this restriction within uncertainty limitations and these must be the states listed above under item 3. If the collision-complex states are chosen as zero-order, then the perturbation is much larger and the collision-complex states are not generally consistent themselves with the Franck-Condon restriction. It is not necessary to examine the entropy restrictions on the activated-complex states since the Franck-Condon restriction requires entropy degeneracy as well as an energy degeneracy.

Small mixing perturbations can be due to solvent motions or to ligand motions. The latter are distortions such as deformations of the ligand shells beyond limits allowed by normal vibrational motion. These are not stationary-state motions and can produce both kinetic-energy and potential-energy perturbations of the electronic states of the reactants. Although these coordinate motions occur near turning points for such motion, they may be sufficiently rapid to be effective. There would not appear to be other motions of the reactants proper which could be effective.

The solvent motions are of several types, including the making and breaking of hydrogen bonds, migration of protons across hydrogen bonds, twisting motions, etc. The migration of protons across hydrogen bonds is an interesting possibility since it is faster than other possible nuclear motions. Whether or not it might be sufficiently large is an interesting question, but large charge displacements are associated with this motion. The perturbation motions must be associated with significant electrical displacements. Thus for water we are logically led to consider the dielectric relaxation behavior. The latter is characterized by a major relaxation process with characteristic time of about 5×10^{-11} sec. and a smaller but faster process with lifetime of about 10^{-13} sec. There is some agreement about the molecular motion involved in the slower perturbation, and it may be possible to write reasonably accurate perturbation operators for this motion even at the present time. In any event, these motions are attractive possibilities for the effective perturbation we seek. In fact, Levich and Dogonadze,[13] in the only serious treatment of non-adiabatic electron-transfer reactions in solution which has appeared, consider only the solvent perturbations. We shall discuss this theory in some detail, but it must be noted that there is as yet no information which allows us to make a choice among those possibilities we have mentioned.   Although all perturbations are dependent on nuclear motions and can be treated within the same theoretical framework it is necessary, as Levich and Dogonadze found, to make quite specific choices before specific expressions for the transmission coefficient can be developed.

The Simple Variation-of-Constants Method

Several reasonable limitations will be placed on the models to be considered. The first of these is that we shall treat the activated complex as a single molecular species with fixed nuclear geometry* having only two important electronic states, namely, that of the reactants' activated complex and that of the products' activated complex. By "important" we mean that all other electronic states lie so much higher in energy as to be of negligible importance. We recognize these states as the stationary states of the activated complex, and they are always a good zero-order approximation since our definition of non-adiabaticity requires that the mixing of these states, upon which the non-stationary behavior depends, be very small. Any mixing perturbations due to nuclear motions can be introduced as small effects. The activated complexes can usually be considered to contain solvent, but it is not always clear how the distinction between solvent thus included and external or bulk solvent is to be made.

In fact there are many pairs of suitable degenerate activated-complex states, not only because there are many possible values of R for which electron transfer will be significant, but also because there are a number of solvent arrangements for a given R which satisfy the Franck-Condon restrictions.   This is also true for variations in nuclear quantum numbers of the reactants proper.   All such states can be divided into pairs such that each pair is identical in nuclear description, but one has the migrating electron on the donor reactant and one on the acceptor reactant.   For complex molecules there are no problems about crossing of potential surfaces, though the mixing which allows electron transfer may be restricted or forbidden by symmetry and momentum restrictions associated with the electronic parts of the wave functions.   An adequate theory must include all these states and all selection rules at some point, but we can nevertheless speak of a "single" reactants' activated complex and a "single" products' activated complex if we remember that the crossing point is really a complex multidimensional crossing surface characterized by a spread of energy and entropy values.

Unless there are very severe symmetry or multiplicity restrictions, inner-sphere reactions should be adiabatic.   Weaker interaction energies will

---

* If the interreactant distance is fixed, the kinetic energy in this coordinate is effectively zero.   This is not, of course, the true situation, and a special procedure must be provided to include this kinetic energy in the energy balance.   Similar remarks apply to the configurational entropy associated with this coordinate.

occur, if at all, primarily in outer-sphere reactions. It is customary also to assume spherical symmetry of the reactants in order to simplify calculations. This is not a very good approximation for complex ion reactants with a heterogeneous ligand composition nor is it very good for organic electron-transfer reactions. Formally, we need make no such assumption.

According to well-known results of the application of the variation-of-constants method, the lower-lying first-order wave function, $\Psi'(t)$, resulting from mixing of the zero-order functions, Ψ'_R and Ψ'_P, is given by

$$\Psi'(t) = [c_R(t)\Psi'_R + c_P(t)\Psi'_P]e^{-iE_0 t/\hbar} \tag{5-2}$$

in which

$$\begin{aligned}\Psi'_R &= \Psi'_{R,v}\Psi'_{R,r}\Psi'_{R,el} \\ \Psi'_P &= \Psi'_{P,v}\Psi'_{P,r}\Psi'_{P,el}\end{aligned} \tag{5-3}$$

with zero-order energy $E_R^0 = E_P^0 = E_0$.

$\Psi'_{R,v}$, $\Psi'_{R,r}$, and $\Psi'_{R,el}$ represent vibrational, rotational, and electronic wave functions, respectively, of the reactants' activated complex. The wave functions with a P subscript represent similar wave functions of the products' activated complex. The probability of finding the electron in the configuration of the reactants' activated complex is

$$|c_R|^2 = \cos^2\left(\frac{H'_{RP}t}{\hbar}\right)$$

Similarly, that for the products' activated complex is

$$|c_P|^2 = \sin^2\left(\frac{H'_{RP}t}{\hbar}\right)$$

Separability has been assumed though it is probable that H'_{RP} may in some cases depend on vibronic operators, in which case the zero-order functions would have to be written more generally. The Hamiltonian operator consistent with this separation can be written as

$$H = H^\circ + H' \tag{5-4}$$

and the perturbation operator may be separated as follows:*

$$H' = H'_{el} + H'_v + H'_r$$

* This separation may be regarded as a device to simplify the exposition in this section. The operator cannot generally be split in this way and we should have included at least electronic-vibrational interaction operators and perhaps also electronic-rotation operators. It now appears that the former are probably the most important of the interaction operators in condensed-phase electron-transfer reactions. Added sophistication at this point is not called for, however, since we have not explicitly accounted for the solvent and there is little point in doing the job in fragments. For this reason, the treatment of Levich and Dogonadze is included at a later point.

Then the transition matrix element, $H'_{RP} = H'_{PR}$, is

$$H'_{RP} = \langle \Psi_P | \mathsf{H}' | \Psi_R \rangle$$

$$= \langle \Psi_{P,v} | \mathsf{H}'_v | \Psi_{R,v} \rangle \langle \Psi_{P,r} | \Psi_{R,r} \rangle \langle \Psi_{P,el} | \Psi_{R,el} \rangle \qquad (5\text{--}5)$$

$$+ \langle \Psi_{P,v} | \Psi_{R,v} \rangle \langle \Psi_{P,r} | \mathsf{H}'_r | \Psi_{R,r} \rangle \langle \Psi_{P,el} | \Psi_{R,el} \rangle$$

$$+ \langle \Psi_{P,v} | \Psi_{R,v} \rangle \langle \Psi_{P,r} | \Psi_{R,r} \rangle \langle \Psi_{P,el} | \mathsf{H}'_{el} | \Psi_{R,el} \rangle$$

The rotational overlap integrals $\langle \Psi_{P,r} | \Psi_{R,r} \rangle$ are unity since the moments of inertia cannot change during electron transfer. The electronic overlap integral $\langle \Psi_{P,el} | \Psi_{R,el} \rangle$ is zero if $\Psi_{P,el}$ and $\Psi_{R,el}$ are orthogonal. Such othogonality is possible because the nuclear geometry is assumed fixed. This assumption can be retained even if the perturbation treated is the potential energy or kinetic energy of the motion in the solvent cage or a ligand displacement motion. Then

$$H'_{RP} = \langle \Psi_{P,v} | \Psi_{R,v} \rangle \langle \Psi_{P,el} | \mathsf{H}'_{el} | \Psi_{R,el} \rangle \qquad (5\text{--}6)$$

The formal simplicity of Eq. 5–6 is misleading since the practical problem in writing $\Psi_{P,el}$ and $\Psi_{R,el}$ is very great. A reasonable way to obtain approximate expressions for $\Psi_{R,el}$ and $\Psi_{P,el}$ is to use first-order time-independent perturbation theory with the wave functions of the separated reactants or products. Then

$$\mathsf{h}_R = \mathsf{h}^\circ_a(\mathbf{r}_a, \mathbf{r}_i) + \mathsf{h}^\circ_b(\mathbf{r}_b, \mathbf{r}_j) + \mathsf{h}^\circ_e(\mathbf{r}_e, \mathbf{r}_a, \mathbf{r}_i) + \mathsf{h}'(\mathbf{R}, \mathbf{r}_a, \mathbf{r}_b, \mathbf{r}_i, \mathbf{r}_j)$$

$$+ \mathsf{h}''(\mathbf{R}, \mathbf{r}_e, \mathbf{r}_b, \mathbf{r}_j)$$

$$\mathsf{h}_P = \mathsf{h}^\circ_a(\mathbf{r}_a, \mathbf{r}_i) + \mathsf{h}^\circ_b(\mathbf{r}_b, \mathbf{r}_j) + \mathsf{h}^\circ_e(\mathbf{r}_e, \mathbf{r}_b, \mathbf{r}_j) + \mathsf{h}'(\mathbf{R}, \mathbf{r}_a, \mathbf{r}_b, \mathbf{r}_i, \mathbf{r}_j) \qquad (5\text{--}7)$$

$$+ \mathsf{h}''(\mathbf{R}, \mathbf{r}_e, \mathbf{r}_a, \mathbf{r}_i)$$

$$\mathsf{h}_R = \mathsf{h}_P \qquad (5\text{--}8)$$

where

h°_a is the Hamiltonian operator for reactant a without the migrating electron, e, at an infinite distance from reactant b but with the solvent and ligand geometry of the activated complex.

h°_b is defined in parallel fashion to h°_a.

h°_e $(\mathbf{r}_e, \mathbf{r}_a, \mathbf{r}_i)$ contains the additional terms due to electron e on a at infinite separation of a and b.

h°_e $(\mathbf{r}_e, \mathbf{r}_b, \mathbf{r}_j)$ contains the additional terms due to electron e on b at infinite separation of a and b.

h' $(\mathbf{R}, \mathbf{r}_a, \mathbf{r}_b, \mathbf{r}_i, \mathbf{r}_j)$ contains the coulombic perturbation terms at the activated-complex value of $\mathbf{R}$ for the interactions of all particles of a with all particles of b but excluding interactions of e.

h'' $(\mathbf{R}, \mathbf{r}_e, \mathbf{r}_b, \mathbf{r}_j)$ contains the coulombic perturbation terms at the activated-complex value of $\mathbf{R}$ between e and the particles of b.

h'' ($\mathbf{R}, \mathbf{r}_e, \mathbf{r}_a, \mathbf{r}_i$) contains the coulombic perturbation terms at the activated-complex value of $\mathbf{R}$ between e and the particles of a.

(The nuclei of a have generalized position vector $\mathbf{r}_a$, those of b have $\mathbf{r}_b$, and the electron position vectors except for electron e are $\mathbf{r}_i$ on a and $\mathbf{r}_j$ on b. All nuclear coordinates are parameters.)

The infinite-separation functions $(\phi_a \phi_b)_n$ and $(\chi_a \chi_b)_m$ must apply to the hypothetical states in which solvent and ligands have been rearranged in the reactants and in the products, respectively. The energy contributions from the rearrangement process are usually too large to be treated by perturbation theory. That these functions themselves will be difficult to approximate is another matter. Although the total Hamiltonian operators are equal (Eq. 5–8), the zero-order Hamiltonian operators defined in

$$\{h_a^\circ + h_b^\circ + h_e^\circ(r_e, r_a, r_i)\}(\phi_a \phi_b)_n = \varepsilon_n^\circ (\phi_a \phi_b)_n$$
$$\{h_a^\circ + h_b^\circ + h_e^\circ(r_e, r_b, r_j)\}(\chi_a \chi_b)_m = \eta_m^\circ (\chi_a \chi_b)_m \tag{5–9}$$

are not, so that $\Psi_{R,\text{el}}$ and $\Psi_{P,\text{dl}}$ found by this procedure and shown in

$$\Psi_{R,\text{el}} = (\phi_a \phi_b)_0 + \sum_{n \neq 0} \frac{V_{0n}^R (\phi_a \phi_b)_n}{(\varepsilon_0^\circ - \varepsilon_n^\circ)}$$
$$\Psi_{P,\text{el}} = (\chi_a \chi_b)_0 + \sum_{n \neq 0} \frac{V_{0m}^P (\chi_a \chi_b)_m}{(\eta_0^\circ - \eta_m^\circ)} \tag{5–10}$$

$$V_{0n}^R = \langle (\phi_a \phi_b)_n \, | h'(\mathbf{R}, \mathbf{r}_a, \mathbf{r}_b, \mathbf{r}_i, \mathbf{r}_j)| \, (\phi_a \phi_b)_0 \rangle$$
$$\qquad + \langle (\phi_a \phi_b)_n \, | h''(\mathbf{R}, \mathbf{r}_e, \mathbf{r}_b, \mathbf{r}_j)| \, (\phi_a \phi_b)_0 \rangle$$
$$V_{0m}^P = \langle (\chi_a \chi_b)_m \, | h'(\mathbf{R}, \mathbf{r}_a, \mathbf{r}_b, \mathbf{r}_i, \mathbf{r}_j)| \, (\chi_a \chi_b)_0 \rangle$$
$$\qquad + \langle (\chi_a \chi_b)_m \, | h''(\mathbf{R}, \mathbf{r}_e, \mathbf{r}_a, \mathbf{r}_j)| \, (\chi_a \chi_b)_0 \rangle \tag{5–11}$$

are not orthogonal. Any attempt to approximate $\Psi_{R,\text{el}}$ and $\Psi_{P,\text{el}}$ in terms of the reactants or products at infinite separation produces the same embarrassment. This is the post-prior Hamiltonian problem (cf. footnote, page 89). Fortunately, in our reactions the problem is not so embarrassing as it looks, since a consideration of the relationships between $(\phi_a \phi_b)_n$ and $(\chi_a \chi_b)_m$ and between V_{0n}^P and V_{0m}^P following, for example, the general approach detailed in Wu and Ohmura[148] (p. 192) demonstrates that the overlap integral $\langle \Psi_{P,\text{el}} \, | \, \Psi_{R,\text{el}} \rangle$ is very nearly zero. This is physically to be expected since the migrating electron is confined to the region of space about a before transfer and to the region of space about b after transfer. Nevertheless, considering the several complications and the lack of information about the zero-order functions for the separated reactants, it may be more useful to start with the total Hamiltonian operator for the activated complex and approximate as directly as possible the two states Ψ_R and Ψ_P which are desired. Another aspect of this problem appears in the discussion of the theory of Levich and Dogonadze (see page 106).

The vibrational overlap integral, $\langle \Psi_{P,v} | \Psi_{R,v} \rangle$, is a product over individual overlap integrals, one for each vibrational mode of the reactants (or products) and the solvent. Many of these will be unity, since their single-mode wave functions are not altered by electron transfer. Some, however, will be less than unity because of force-constant changes resulting from electron transfer or because there is a change in vibrational quantum number. The formalism of Eq. 5–6 depends on separability of the wave function and is not adequate for a detailed investigation of the changes in vibrational quantum numbers. If energy balance is achieved in this way, changes in quantum numbers for modes not coupled to the migrating electron can take place spontaneously with transfer but with very low probability, so that it may be possible to ignore such processes. On the other hand, quantum-number changes in modes coupled to the migrating electron are much more probable and, since there are a number of different ways in which the energy balance can be secured using such quantum-number changes, they are probably important in electron transfer. Gurnee and Magee[152] have made some estimates of the importance of this kind of process in simple gas-phase electron-transfer reactions and find it significant. The matter has not yet been considered for electron-transfer reactions in solution.

The discussion of electron transfer has been restricted to single-electron migrations. *A priori*, there is nothing in Eq. 5–6 to suggest that simultaneous two-electron migration processes have smaller H'_{RP} and thus smaller transmission coefficients than one-electron processes, and it is now well established that such processes occur. The relative magnitudes H'_{RP} for such processes may be quite different but, if both are sufficiently large, the reaction will be adiabatic and independent of the transmission coefficient. Only calculations of H'_{RP} for both kinds of transfer can tell us whether there is any reason to expect two-electron processes to lead to non-adiabatic reactions more often than not. The Gurnee-Magee work has provided some information on this point though the information applies to gas-phase reactions involving a perturbation by motion in the R coordinate. They found that the probability for simultaneous two-electron migration in the homonuclear reaction

$$Ne + Ne^{+2} \rightarrow Ne^{+2} + Ne$$

as calculated by their method, was only slightly smaller than the one-electron transfer probability for

$$Ne + Ne^{+} \rightarrow Ne^{+} + Ne$$

Apart from some scattering theories based on the method of molecular beams, this is the only theoretical comparison of one-electron and two-electron processes which we have and though it is in qualitative agreement with experimental observations that two-electron processes are not, as a class, slower than one-electron processes, we would be unwise to generalize at this point.   Energy differences between reactants and products are usually greater for two-electron than for one-electron processes, as are geometric differences.   Hence, even though transmission coefficients for simultaneous two-electron transfer may not be much smaller than those for single-electron transfer, the energy-balance problem is more severe, and the reaction likely to be slower.   Often there is also the possibility that electrons move one at a time in two one-electron reactions which, as a consequence, then involve two elementary steps, each with its own activated complex.   Examples of both kinds of behavior are known in electron-transfer reactions, and the relative importance of single-step and double-step processes can be roughly estimated on the basis of the instability of the one-electron intermediate.   It was for some time believed that two-electron processes would be unusually slow. This no longer appears to be the case, and comparisons of similar one-electron and two-electron processes discussed in Chapter 3 suggest that, in general, two-electron processes are just about as rapid as one-electron processes.

The involvement of excited states in electron-transfer reactions has not yet been studied in any detail.   There are several possibilities.   In the first of these, a reaction with a large positive over-all free-energy change may occur via a pathway involving thermal promotion of one or, in very special cases, both reactants to excited electronic states in order to achieve the energy-balance condition.   Such a process is, however, unlikely when compared with thermal excitation of vibrations and vibrational quantum-number changes during electron transfer unless there are low-lying excited states, since over-all free-energy changes in most chemical reactions are small relative to the usual promotion energies to excited states.   In the opposite situation, that is, when the over-all free-energy change is very large and negative, the products may be formed in excited electronic states. Again the probability of such processes depends on the relative energy of the excited states and the over-all free-energy change.   It is to be expected that processes of the latter type, if they occur, will occasionally be found to emit radiation in the de-excitation of the products.   Recently, Hercules[150] and Chandross and Sonntag[151] have observed chemiluminescence in oxidation reactions employing strong oxidizing agents and strong reducing agents.   In the studies by the latter authors, the 9,10-diphenylanthracene negative ion on oxidation by chlorine, for example, emitted light characteristic of 9,10-diphenylanthracene itself.   This is not what would have been expected since, if the highest-lying electron of the ion is transferred to chlorine, chlorine might have become excited as the Cl_2^- ion without anthracene excitation. The authors propose that the electronic energy is released as local heat so rapidly with respect to the collision lifetime in the cage that the anthracene product is thermally excited and then emits. However, they also suggest the interesting possibility that it is not the highest-energy electron of 9,10-diphenylanthracene ion which is abstracted by chlorine, but rather the highest bonding electron of the negative radical ion, to form directly the anthracene product in an excited state. This suggestion appears to be quite reasonable if the energy-balance condition is more readily met with the lower-lying orbital, and quantitative considerations of the energy situations for specific cases should provide a distinction between the suggested alternatives.

The Selection Rules

If a reaction is non-adiabatic, the slow step is the electron-transfer process in the activated complex, which may be treated as an elementary reaction and given a rate constant designated k_{RP}. This rate constant is given by the transmission coefficient, which is equal to the probability of transfer on a single passage through the activated-complex region, P_{RP}, divided by the lifetime of the transition state; or, if there are several vibrations of nuclear coordinates which take the system through the activated-complex state before it decomposes, $k_{RP} = \nu P_{RP}$, in which ν is the number of vibrations per second, i.e., the number of times the perturbation P_{RP} is applied per second. In the latter case, some consideration of the probability of return to the reactants' activated-complex state may be required. However, it is doubtful that reactions of present concern pass through the activated-complex state more than once on any collision. Repeated formations of the activated complex before the reactants diffuse out of the solvent cage require special consideration in adiabatic reactions but not in non-adiabatic reactions.

In view of what has been said about the times of solvent and ligand motions, we can conclude that the nuclear motions which destroy an activated complex for electron transfer in ordinary solvents at room temperature are probably effective in times no longer than 10^{-12} sec. Hence $k_{RP} \leqslant 10^{12}$ sec.$^{-1}$ is required if the reaction is to be classed as non-adiabatic. A simple uncertainty-principle argument shows that this limit is equivalent to a maximum interaction energy of ± 100 cal./mole and is consistent with the expectation discussed in Chapter 1. Such small values of H'_{RP} would seem probable only in cases where there are very effective restrictions on the matrix element of the internuclear coulombic operators. Such restrictions normally occur because of failures to conserve parity or momentum in a process and are responsible for the well-known selection rules of spectroscopy. The selections for predissociation are a guide to those for electron-transfer reactions, since predissociation as a crossing process between states of equal energy involves similar momentum restrictions. The parity rule will depend on the inversion behavior of the interaction operator. The exact formulation of the parity rule for homonuclear electron-transfer reactions is complicated by the degeneracy of the activated-complex state and by the transferring electron. However, activated complexes can be constructed in a number of different ways using different vibrational quantum numbers, and unless the reactants are especially simple there will be no problem with parity conservation. As a result, the only selection rules which need be considered are those based on conservation of angular momentum.

In the conservation of angular momentum, it is unlikely that the solvent in the second and outer solvation spheres can play any significant role. During the time of electron transfer, which is of the order of 10^{-15} sec., the momentum must be conserved as the system makes the non-adiabatic transition between the zero-order surface of the reactants' activated complex and the zero-order surface of the products' activated complex. Because of the weak coupling between the electrons in the reactants and the solvent in the outer solvation spheres, transfer of momentum from solvent nuclei to the electrons of the reactants during this very short period of time is highly improbable. Even the electronic polarization of the solvent which can follow the change in charge distribution on the reactants is not useful for momentum conservation because there is no ready mechanism for transfer of momentum between electrons in the solvent molecules and electrons in the reactants.* Therefore, we conclude that the outer solvent is not useful for conservation of momentum and that, for non-adiabatic reactions, the momentum selection rules apply to the reactants proper. Furthermore, we note that regardless of the perturbation the orbital angular momentum of the zero-order states must be the same, i.e., the states must be of the same symmetry. If the zero-order states belong to different irreducible representations, they cannot be mixed. With a strong interaction between reactants, the orbital angular momenta couple, and the products' activated complex must have one of the orbital angular-momentum quantum numbers obtained by vectorial combination of the total angular momenta of the separate reactants. If the interaction is weak, the total orbital angular momentum of each reactant must be conserved independently after correction is made for the migrating electron. However, momentum exchange between electrons and nuclei of the ligands in complex ions of the first transition series is very efficient; of course, it is far less so for complexes of the rare earth ions. If the change of total electronic angular-momentum quantum number is such that orbital momentum conservation cannot be established within the electron system, it is required that momentum be exchanged with the nuclei efficiently during the lifetime of the activated

---

* The momentum exchange at the time of electron migration between the activated complex and the mode of motion responsible for the mixing perturbation is not zero, but it is very small and of widely varying amount. This momentum and the energy exchange associated with the perturbation can be ignored.

complex. Good estimates of these times are not available, but they must be very fast since it is the electronic motional times which limit the process. Furthermore, there are many possibilities for electron-nuclei momentum exchange. Hence it is probable that selection rules based on orbital angular-momentum conservation are usually unimportant in electron-exchange reactions of first-transition series complexes. Such rules may be more restrictive in the electron-transfer reactions of organic molecules.

In general, if there are symmetry restrictions on electron transfer through the activated complex lying at lowest free energy, there will be alternate pathways through higher-lying activated complexes for which no symmetry restrictions exist. The latter activated complexes need not be reached by outright promotion but rather by relatively inexpensive distortion of the framework of the reactants. Such vibronic interaction which decreases electronic symmetry in an appropriate fashion will almost always provide sufficient mixing of ground and excited states to minimize or eliminate symmetry restrictions. The lower-lying pathway would be characterized by a low activation energy but a large negative value of the apparent entropy of activation; the alternate pathways by higher activation energy and less negative values of the activation entropy. It is probable that suitable nuclear distortion can usually occur such that one of the alternate pathways provides a higher net rate of reaction than the symmetry-restricted pathway. Similarly, when one of the reactants has a Jahn-Teller distortion which it loses on electron transfer or if the dimensions or coordination number change in the process, the activated complex will have a geometry lying somewhere between the normal equilibrium geometries of reactants and products. Such distortions will tend to minimize symmetry restrictions and will, of course, also increase the activation energy.

The situation is quantitatively different and probably qualitatively different for spin momentum conservation. If we first ignore momentum exchange between spin systems and orbital electron systems, the Wigner-Wittmer[153] rules apply for a strong spin-spin interaction between reactants. Then if S_a and S_b are the spin quantum numbers of the two reactants in the reactants' activated complex, the total spin quantum number of the products' activated complex must have one of the values $S_a + S_b$, $S_a + S_b - 1$, $\ldots$, $|S_a - S_b|$. If the interaction between spin systems of the two reactants is weak, as it may often be for outer-sphere reactions with small delocalization onto the ligands, the spin angular momentum aside from that associated with the migrating electron must be conserved on the separate reactants. It is now necessary to decide whether or not these spin restrictions are likely to be important in electron-transfer reactions.

In many crossing processes or intersystem changes, the multiplicity selection rule is not very effective. Thus despite the weak spin-orbital coupling in benzene, the intersystem crossing from the first excited singlet state to the ground triplet state is efficient and is, in fact, a first-order process in the gas phase so that collisional processes are unnecessary. This appears to be a consequence of the relatively long time during which the resonance conditions exists. In electron-transfer reactions very much shorter times are encountered, so that the spin-orbital interaction, upon which momentum transfer into or out of the spin system depends, must be more efficient, that is, it must be sufficiently large so that spin momentum can be changed in 10^{-12} to 10^{-13} sec. or perhaps even faster. It is true that large atoms have increasingly large spin-orbital interaction, but we also know from direct and indirect measurements of the spin-lattice relaxation times for the spin systems of some first-transition series complexes that a lower limit for the time of this process is about 10^{-12} sec. and that longer times are common. Thus for the hexaaquoiron(III) complex, the time is about 10^{-11} sec., and for manganese complexes it may be as long as 10^{-6} sec. These times are specifically the times required for momentum exchange between spin systems and the rest of the world, and they thus measure the probability that the spin selection rules just described can be violated. On the basis of this consideration, we are forced to conclude that the spin selection rules are probably of major importance in electron-transfer reaction. Indeed, it is probable that if there are non-adiabatic reactions between complex ions, they must be due to violations of the spin selection rules. Even for inner-sphere reactions, these selection rules may be important.

The importance of spin conservation in the electron-transfer reactions of molecules consisting of small atoms only, as for example most organic molecules, is a bit easier to estimate since well-known observations like the long lifetimes of the ground triplet states of these molecules make it reasonably certain that at least the Wigner-Wittner rules must be adhered to. Other observations indicate, however, that when these rules are met, other momentum selection rules are minor. The efficiency of triplet-state quenching by singlet-state quenchers with spin conservation demonstrates that there is good coupling of the spin systems of the reactants at kinetic-theory collision distances.

Returning to complex ions, reaction (5–1) has already been mentioned as a possible case in point. The arguments of the present section suggest that its unusual slowness is attributable to the large change in spin quantum number and not to some other cause. Another well-known example of a possible spin restriction is the oxidation of hemoglobin by oxygen. Such evidence as is available[154,155,156] indicates that the oxidation of the Fe(II) form of this protein occurs most rapidly when water rather than oxygen is bound directly to the iron ion. Electron transfer then must occur through the porphyrin ring of the heme group or through an unsaturated pathway elsewhere in the protein. The oxygen compound has zero magnetic susceptibility but the water compound of the Fe(III) species have maximum uncoupled spin systems. There is thus a large spin restriction for the oxidation of the oxy(II) form and, depending on the oxygen product, a small or zero restriction for oxidation of the aquo(II) form. However, studies of the oxidation of hemoglobin and related heme-proteins by ferricyanide ion[157,158] show that the situation is not this simple and that differences in rearrangement free energies must be at least as important as differences in spin states. It is probable in iron porphyrins that low-spin and high-spin species have very different equilibrium geometries so that there is a large activation energy for any electron-transfer reaction which converts a low-spin iron-porphyrin complex into a high-spin complex or vice versa. This is especially true for the heme-proteins, in which some accommodation between ligand geometry and protein geometry must be established before and after electron transfer. In these proteins, geometric restrictions resulting from the structural requirements of the protein can be expected to dominate the electron-transfer process just as they probably determine the equilibrium electronic properties of the complexed iron ion. Thus there are several reasons why the oxidation of hemoglobin and myoglobin would be expected to be slow. On the other hand, both oxidized and reduced states of the heme-protein cytochrome c are spin-paired. There is no spin restriction and only a small ligand rearrangement is required, so that oxidation should be rapid, as indeed Sutin[158] has found.

It is worth noting that nuclear distortions of the type which can eliminate symmetry restrictions on electron transfer would not appear to have any such effect when spin restrictions are involved. It is necessary that nuclear distortions considerably increase the spin-orbital coupling if spin restrictions are to be decreased, but the effect of nuclear distortions on spin-orbital coupling is a distinctly second-order effect and thus unlikely to be of sufficient magnitude.

The Treatment of Zener, Landau and Stueckelberg

It is clear that the elementary application of the variation-of-constants method generally used to discuss resonance in the presence of a steady perturbation, as is done in elementary treatments of benzene, is inapplicable to the electron-transfer reactions in which the zero-order states of the ac-tivated complex and the perturbation operator are all functions of time. A more complicated application of the variation-of-constants method, the perturbed stationary-state method, or the distorted wave method is required. The expression for the transmission coefficient (transition probability) used most often for non-adiabatic chemical processes is that of Zener,[147] Landau,[146,159] and Stueckelberg.[160] All three men arrived at the same final result at about the same time, and all used different methods although Stueckelberg's treatment is the most extensive as well as the most rigorous. The simplified result for the transition probability, P_{RP}, between the zero-order energy surfaces E_R and E_P of Fig. 5–1, is given by

$$P_{RP} = 1 - \exp\left(-\frac{4\pi^2 |H'_{RP}|^2}{hv |\lambda_R - \lambda_P|}\right) \simeq \frac{4\pi^2 |H'_{RP}|^2}{hv |\lambda_R - \lambda_P|} \qquad (5–12)$$

where H'_{RP} is evaluated at the crossing point of the zero-order surfaces, λ_R and λ_P are the derivatives of the zero-order surfaces with respect to the reaction coordinate, as shown in Fig. 5–1, and v is the average velocity in the reaction coordinate at the crossing point. The λ's enter the problem as forces at the crossing point. The matrix element H'_{RP} is obtained by integration over the electronic coordinates at fixed nuclear coordinates, R_0, of the activated complex. Since there are many sets of acceptable nuclear configurations for activated complexes, H'_{RP} would have to be evaluated for each set of nuclear configurations in a more detailed treatment. In general, different nuclear configurations would give different H'_{RP} values, different velocities, different λ's, and different rearrangement free energies. Note that P_{RP} as given in Eq. 5–12 is independent of time.

All three treatments make essentially the same important assumptions which seriously limit application of the result. Equation 5–12 has been used almost without discrimination in a wide variety of non-adiabatic problems from electron-transfer to predissociation and solid-state phenomena. The appropriateness of some of these applications may be questioned. In particular, Coulson and Zalewski[161] have suggested that it may be poor for any process with a small velocity along the reaction coordinate and hence perhaps poor for all electron-transfer reactions in solution that are "slow" even with the fastest nuclear perturbation possible at ordinary temperatures.

It should also be observed that the several treatments implicitly rest on a model of the crossing-point region no more complicated than that for a diatomic molecule. Actually, for polyatomic reactants the "crossing point" is a multidimensional surface in phase space about which very little is known. Certainly the size is vague, and there can be more than one reaction coordinate for the same over-all reaction even if these coordinates are all complicated functions of a common parametric distance.

A detailed discussion of the assumptions, approximations, and limitations of the simple equation is given by Coulson and Zalewski,[161] Bates,[162] and Marcus.[19] As yet, no substitute expression has been provided. However, Nikitin[163] has modified the original Landau treatment in a direction useful for our purposes but otherwise still unsatisfactory since it retains many of the approximations. The original treatments are limited by the requirement that the relative kinetic energy in the reaction coordinate be very much larger than H'_{RP}. We cannot expect this restriction to be generally appropriate to liquid-phase reactions and, particularly, it cannot be appropriate for perturbing motions near turning points of these motions. Even though the interaction matrix elements we are considering are very small, the relative kinetic energies in some types of reaction coordinates can also be small. Thus it would appear that Nikitin's treatment, which removed this restriction, is preferable to the original. His result is given by

$$P_{RP} = \frac{4\pi \, |H'_{RP}|^2}{\hbar} \left(\frac{\mu}{f \cdot \Delta f} \right)^{2/3} \tag{5-13}$$

in which f is the average force in the reaction-coordinate direction at the crossing point, Δf is $|\lambda_R - \lambda_P|$, and μ is the effective mass for this coordinate.

The confidence engendered by Eq. 5–12 which results in its almost universal use is often misplaced, especially when applied to electron-transfer reactions. Reasonable orders of magnitude for P_{RP} can probably be obtained by insertion of a consistent set of parameters but many other even more approximate expressions can be used to give order-of-magnitude estimates of P_{RP}. The principal problem in using Eq. 5–12 is the lack of knowledge of the source of the perturbation, the strength of its coupling to the activated complex, and the dynamics of the perturbation as well as those of the reaction system along the reaction coordinate. However, it can be shown by a rather simple argument that since it is the rate constant, k_{RP}, for the electron-transfer act treated as an elementary reaction which is desired, most of the confusion associated with estimation of the dynamical parameters can be avoided with little if any loss in the over-all estimate of $P_{RP} \nu_{RP} = k_{RP}$, ν_{RP} being in this treatment the reciprocal of the lifetime of the activated complex, τ. We have already estimated this time on the basis of a very limited knowledge of the nuclear motional times. The problem of estimating τ can be approached in other ways. For example, Kauzmann[164] has taken τ to be the length of time in which the activated complex has an energy in the range $E \pm 4 \, H'_{RP}$ where $E = E_R = E_P$ at the crossing point of the zero-order surfaces. Outside of this range, the energies of the zero-order and first-order surfaces differ by less than 10 percent. This arbitrary choice certainly does not include all the distance along the reaction coordinate in which the probability of electron transfer is at all important but is sufficient for the present approximation. The expression obtained for τ was

$$\tau \simeq \frac{8 \, |H'_{RP}|}{v \, |\lambda_R - \lambda_P|}$$

Combining this with Eq. 5–12, it is readily seen that

$$k_{RP} \simeq \frac{\pi^2 \, |H'_{RP}|}{2h}$$

that is, that k_{RP} is directly proportional to $|H'_{RP}|$ when the latter is very small. This estimate has the virtue that it does not depend on mechanical and temporal details of motion in the reaction coordinate. If the over-all reaction is simplified thus:

$$D + A \underset{k_{-1}}{\overset{k_1}{\rightleftharpoons}} X_R^{\ddagger} \underset{k'_{RP}}{\overset{k_{RP}}{\rightleftharpoons}} X_P^{\ddagger} \underset{k_{-3}}{\overset{k_3}{\rightleftharpoons}} D' + A'$$

the reaction can only be non-adiabatic if $k_{-1} > k_{RP}$ and $k_3 > k'_{RP}$. The diffusion processes k_1 and k_{-3} are unimportant. The constants k_{-1} and k_3 are reciprocals of the lifetime, τ, of the activated complex. Hence, we again find a limit to $|H'_{RP}|$ of no more than 100 cal./mole; the reaction would still be partially adiabatic at somewhat lower interaction energies. If a value of τ less than 10^{-12} sec. is chosen, then $|H'_{RP}|$ must be correspondingly less. This reasoning supports the skepticism as to the occurrence of non-adiabatic reactions expressed at the beginning of this chapter.

Thus far, it has been unnecessary to distinguish between resonance processes, in which the electronic energy of the reactants is identical within uncertainty limits with that of the products, and processes in which energy is conserved in combinations of electronic and vibrational coordinates. None of the discussion has made it necessary to consider separate matrix elements of vibrational operators even if there are changes in vibrational quantum numbers on electron transfer. There is some question about the generality of this neglect of vibration factors in the matrix elements. If any vibrational mode. changing quantum numbers participates in H'_{RP}, then the Franck-Condon factors would appear to be adequate. On the other hand, if such a mode is not coupled to the electronic transition as, for example, when it is independent of the interreactant distance R and the perturbation, there should be an additional factor in P_{RP} for the probability of change of quantum number in this isolated mode. Obviously, this factor will be considerably less than unity, so that a process of this type will have a smaller k_{RP} than one involving a participating mode or no vibrational quantum-number changes at all. It is interesting to note (in anticipation of Chapter 6) that current theories of electron-transfer reactions do not include energy balance by vibrational quantum-number changes. At present, energy balance is split between electronic states (including solvation) and vibrational states with rearrangement to the degenerate (in total energy) activated complex occurring in both kinds of coordinate. It is not clear why vibrational quantum-number changes are excluded. So long as the reaction is adiabatic, energy balance achieved this way is important since the mechanism provides many different ways to achieve this necessary condition. Even for non-adiabatic processes, vibrational quantum-number changes may be important. The individual Franck-Condon vibrational factors for such processes are smaller, though they may not be much smaller, since large force-constant changes with electron transfer occur in many modes to reduce these factors. There are additional activated complexes possible when vibrational quantum numbers can change.

To avoid confusion, it should be noted that entropy considerations do not restrict the manner in which the energy balance in the activated complex is obtained. It is only necessary that the entropy itself be unchanged on electron transfer. The conservation of entropy has been stressed by Marcus, and it is a direct consequence of the Franck-Condon principle just as is the conservation of energy within the activated complex. Strictly speaking, the activated complex is not totally isolated from its environment with respect to energy and entropy exchange with the environment, since electronic polarization interactions with the environment are always adiabatic with respect to changes in the electronic quantum states of the activated complex. An obvious example of this is the introduction of free energy from a radiation field. The adiabatic behavior of electronic polarization is a feature of the theories discussed in Chapter 6, and the small contributions to entropy and energy balance are thus accounted for. Hence in order to apply the Franck-Condon principle without confusion, the activated complex should be defined to include interactions with solvent and solvent ions.

The Theory of Levich and Dogonadze

The most extensively applied theories of electron-transfer reactions are those of Marcus and Hush to be discussed in Chapter 6. These theories were developed for adiabatic reactions but can be modified for application to non-adiabatic reactions by simple changes and the incorporation of an electron-transfer rate constant k_{RP} to replace a collision number. McConnell[14] has treated the non-adiabatic case in an interesting, comprehensive way which does not make such a distinct separation between the transmission coefficient and the free energy of activation. His theory was designed for and applied to electron transfer between conjugated organic ring structures linked to each other through straight-chain coupling links. Instead of discussing his method, which has somewhat limited applicability, it will be more illuminating to present a brief description of the theory of Levich and Dogonadze,[13,165] which is broadly based and includes both adiabatic and non-adiabatic reactions though it is essentially a non-adiabatic theory. It provides an interesting contrast to the theories of Marcus and Hush though the results, which will be compared at a later point, are not so different as might be expected from the difference in model.

Levich and Dogonadze use a macroscopic continuum model for the solvent, starting immediately at the surface of the reactants proper, and they ignore nuclear rearrangement of the reactants themselves. Nuclear rearrangement is important is most electron-transfer reactions but can be formally included in the theory without any fundamental change in the approach. The authors assume that the perturbation mixing the states of the activated complex is due to fast solvent motions, such as the breaking of hydrogen bonds, torsional motion across a hydrogen bond, etc., as discussed at the beginning of this chapter. The time-independent parts of the total Hamiltonian can be ignored. The remaining terms are

$$H(\mathbf{r}, q) = H_e(\mathbf{r}) + H_s(q) + V_{es}(\mathbf{r}, q) \qquad (5\text{--}14)$$

in which $H_e(\mathbf{r})$ is the Hamiltonian of the migrating electron without V_{es}, $H_s(q)$ is the Hamiltonian of the solvent, and $V_{es}(\mathbf{r}, q)$ is the electron potential energy resulting from the polarization induced by the electron in the solvent. $H_e(\mathbf{r})$ is given by

$$H_e(\mathbf{r}) = -\frac{\hbar^2}{2m}\,\nabla_r^2 + U_a(\mathbf{r}, R) + U_b(\mathbf{r}, R) \qquad (5\text{--}15)$$

in which $(-\hbar^2/2m)\nabla_r^2$ is the kinetic energy of the electron, $U_a(\mathbf{r}, R)$ is the interaction between the electron and the reactant a plus the solvent statically polarized by the reactant, and $U_b(\mathbf{r}, R)$ corresponds to U_a for reactant b. The reactant nuclear coordinates are fixed at R_0. $H_s(q)$ is given by

$$H_s = \frac{2\pi}{c}\int \left(|\mathbf{P}_i|^2 + \frac{1}{\omega_0^2}\,|\dot{\mathbf{P}}_i|^2\right) d\tau \qquad (5\text{--}16)$$

in which $\mathbf{P}_i$ is the polarization of the solvent due to the electron but includes atomic and orientation polarization only, since the optical polarization of the solvent is always in phase with the electron. The fundamental solvent parameter is

$$c = \frac{1}{\varepsilon_{op}} - \frac{1}{\varepsilon}$$

in which ε_{op} is the square of the refractive index of the solvent and ε is the low-frequency dielectric constant of the solvent. The motional spectrum of the solvent is expanded in a Fourier series using the solvent coordinates, each term of which has associated a characteristic angular frequency. There are, however, a limited number of these frequencies. The motions are approximated by harmonic oscillator motions so that $H(q)$ is the time-dependent polarization energy of the solvent. This method of expressing polarization modes as Fourier waves is now quite common. It is called a polaron treatment and is due primarily to Pekar.[166] It is at this point that the theory could be extended to include ligand vibrations by adding terms for these motions. However, useful solutions can be obtained only if all but one such motion is eliminated, say ω_0, which is related to the characteristic time of this perturbing motion by $2\pi/\omega_0$. In this case the non-electronic dynamic polarization $\mathbf{P}_i$ is a continuous function of the coordinates and the problem becomes tractable. Then

$$H_s(q) = \frac{1}{2}\sum_k \hbar\omega_0\left(q_k^2 - \frac{\partial^2}{\partial q_k^2}\right) \qquad (5\text{--}17)$$

The rearrangement step to form the activated complex is now necessary. As usual, let us distinguish a collision complex of the reactants with solvent in equilibrium with the charge distribution of this reactants' collision complex. If the electron were to move to produce the products with this same nuclear geometry of solvent and reactants, the energy, in general, would not be conserved (Franck-Condon principle). Hence the equilibrium distribution of solvent is not the one suitable for the degenerate activated complex. The nuclei of the reactants and solvent must undergo rearrangement to some state not in equilibrium with the charge distribution of the collision complex but which does establish the energy degeneracy. This is the final step in forming the activated complex. The required rearrangement of solvent is expressed as a Taylor's series about the equilibrium q_k^0 values of the reactants' or products' collision complexes.

The term V_{es} contains three contributions: (1) electron interaction with solvent polarization induced by the electron itself, (2) electron-polaron interaction, and (3) a correction term resulting from anharmonicity in solvent atom vibrations, i.e., a correction to electron-polaron interaction.

Slow solvent motions of the type responsible for diffusion are considered too slow to be of concern, an assumption to be discussed shortly. The fast vibrational motions are also slow compared with times for electronic motion so the wave function of the system is approximated by the Born-Oppenheimer separation; thus

$$\Psi'_n(\mathbf{r}, q) = \sum_r \Phi_{rn}(q) v_r(\mathbf{r}, q) \tag{5-18}$$

As usual, all electronic states other than the ground state of the reactants' activated complex and that of the products' activated complex are assumed to be too high in energy to be important. The two low-lying states are jointly designated r or r' or individually with subscripts R and P. There are many solvent states n. The function v_r is either Ψ'_R or Ψ'_P determined by

$$\left[\frac{-\hbar^2}{2m} \nabla_r^2 + U_a + V_{es}\right]\Psi'_R = \varepsilon_R(q)\Psi'_R(\mathbf{r}, q) \tag{5-19a}$$

$$\left[\frac{-\hbar^2}{2m} \nabla_r^2 + U_b + V_{es}\right]\Psi'_P = \varepsilon_P(q)\Psi'_P(\mathbf{r}, q) \tag{5-19b}$$

The function $\Psi'_n(\mathbf{r}, q)$ is an eigenfunction of the total time-dependent Hamiltonian with eigenvalue E_n, and on application of this Hamiltonian to Ψ'_n we find

$$\{H_s(q) + \varepsilon_r(q) - E_n\}\Phi_{rn}(q) =$$
$$= \sum_{r'} \{L_{rr'}(q) - \Delta_{rr'}(q)[H_s(q) + \varepsilon_{r'}(q) - E_n]\}\Phi_{r'n}(q) \tag{5-20}$$

In Eq. 5-20, r' is a dummy index for summing over the states r. $E_n - \varepsilon_r(q)$ is the solvent energy. The $\Phi_{rn}(q)$ are not the eigenfunctions of H_s but are the correct solvent wave functions consistent with the arbitrary choice upon which Eq. 5–18 rests. The quantity $\Delta_{rr'}$ is an electronic overlap integral and cannot be responsible for mixing in a first-order perturbation; it is also assumed small at large R_0.

The symbol $L_{rr'}(q) \equiv L_{rr'}^{(1)} + L_{rr'}^{(2)}$ is given by

$$L_{rr'} = -\int v_r^*[H_s, v_{r'}]\, d\mathbf{r} - \int v_r^* U_{r'}\, v_{r'}\, d\mathbf{r} \tag{5–21}$$

$$U_{r'} = U_a \quad \text{if } r' = a; \qquad U_{r'} = U_b \quad \text{if } r' = b$$

where $\lambda[H_s, v_{r'}]$ is the commutator. The first integral on the right of Eq. 5–21 is called the non-adiabaticity operator. The second term is an exchange integral. Both have been examined by Levich and Dogonadze, but the correct choice of either or rejection of both as the perturbation operator can only be determined in a given case by prior detailed analysis of solvent motions.

In any event, if R_0 is sufficiently large, the right-hand side of Eq. 5–20 may be set equal to zero to give the approximations to Φ_{rn}, Φ_{rn}^0 of Eqs. 5–22.

$$\{H_s(q) + \varepsilon_R(q) - E_{Rn}^0\}\Phi_{Rn}^0 = 0 \tag{5–22a}$$

$$\{H_s(q) + \varepsilon_P(q) - E_{Pn'}^0\}\Phi_{Pn'}^0 = 0 \tag{5–22b}$$

But is it necessary now to include the non-equilibrium reorientation of the solvent about equilibrium positions of the solvent coordinates q^0. This involves adding the distortion energy of the solvent oscillators, up to the first power in q, to the energy $\varepsilon_r(q^0)$ so that

$$\varepsilon_r(q) = \varepsilon_r(q^0) + \sum_k V_{kr}(q_k - q_{kr}^0) \tag{5–23}$$

If, however, the force constant V_{kr} is taken to be the normal force constant at coordinate extension q_{kr}^0, considerable simplification is achieved and Eq. 5–22 becomes Eq. 5–24. This choice of V_{kr} is subject to some question when applied to nearby solvent and is almost certainly poor if applied to non-equilibrium distortions of the complex–ion ligand vibrations.

The zero-order wave functions to be used in the perturbation calculation of P_{RP} are the solutions, Φ_{Rn}^0 and $\Phi_{Pn'}^0$ of

$$\left\{H_s(q - q^0) + \varepsilon_R(q^0) + \sum_k \hbar\omega_0 q_{kR}^{02} - E_{Rn}^0\right\}\Phi_{Rn}^0 = 0 \tag{5–24a}$$

$$\left\{H_s(q - q^0) - \varepsilon_P(q^0) + \sum_k \hbar\omega_0 q_{kP}^2 - E_{Pn'}^0\right\}\Phi_{Pn'}^0 = 0 \tag{5–24b}$$

and have zero-order energies E_{rn}^0 given by

$$E_{Rn}^0 = \varepsilon_R(q^0) + \sum_k \omega_\gamma q_{kR}^{02} \tag{5–25a}$$

$$E_{Pn}^0 = \varepsilon_P(q^0) + \sum_k \omega_0 q_{kP}^{02} \tag{5–25b}$$

The total energies of the states, exclusive of the energies necessary to form the reactants' or products' collision complexes, are the $E_{rn}^{0'}$ given by

$$E_{Rn'}^0 = E_{Rn}^0 + \sum_k \hbar\omega_0(n_k + \tfrac{1}{2}) \tag{5-26a}$$

$$E_{Pn'}^{0'} = E_{Rn'}^0 + \sum_k \hbar\omega_0(n_k + \tfrac{1}{2}) \tag{5-26b}$$

These now include the stationary polaron energy of the solvent, with n_k the quantum number of the kth solvent oscillator. The energy $\varepsilon_R(q^0)$ is the total energy of the migrating electron in the field of the reactant ions and the solvent polarization due to these ions and the electron itself. The sum $\sum_k \omega_0 q_{kR}^{02}$ is the energy of the solvent due to polarization by the electron.

If now the perturbation operator is generally designated L_{PR}, $k_{RP} = P_{RP}(\omega_0/2\pi)$ and is given by

$$k_{RP} = \frac{2\pi}{\hbar} A(n) \sum_{n'} |\langle \Phi_{Rn'}^0 | \mathsf{L}_{PR} | \Phi_{Rn}^0 \rangle|^2 \delta(E_{Pn'}^0 - E_{Rn}^0) \tag{5-27}$$

Because of the adiabatic separation of solvent and electronic modes and because the remaining nuclear coordinates are fixed, the matrix element is a function of electronic coordinates only at parametric values of the solvent nuclear coordinates and can be assumed to be evaluated as some suitable average set of these coordinates. The factor $A(n)$ is the statistical average over all initial polaron states n. The sum over n' would include all final polaron states regardless of energy, but only those satisfying the energy-balance requirement are physically acceptable. The δ functions are thus necessary and are functionally expressed in terms of Fourier functions, using plane waves, as in

$$k_{RP} = \frac{1}{\hbar^2} |L_{PR}|^2 A(n) \int_{-\infty}^{\infty} \sum_{n'} \langle \Phi_{Rn}^0 | e^{iE_{Pn}^0 t/\hbar} | \Phi_{Pn'}^0 \rangle$$
$$\times \langle \Phi_{Pn'}^0 | e^{-iE_{Rn}^0 t/\hbar} | \Phi_{Rn}^0 \rangle \, dt \tag{5-28}$$

or

$$k_{RP} = \frac{1}{\hbar^2} |L_{Pr}|^2 A(n) \int_{-\infty}^{\infty} \sum_{n'} \langle \Phi_{Rn}^0 | e^{iH_{sP}t/\hbar} | \Phi_{Pn'}^0 \rangle$$
$$\times \langle \Phi_{Pn'}^0 | e^{-iH_{sR}t/\hbar} | \Phi_{Rn}^0 \rangle \, dt \tag{5-29}$$

On summing over n', Eq. 5–29 becomes

$$k_{RP} = \frac{1}{\hbar^2} |L_{PR}|^2 A(n) \int_{-\infty}^{\infty} \langle \Phi_{Rn}^0 | e^{i(H_{sP} - H_{sR})t/\hbar} | \Phi_{Rn}^0 \rangle \, dt \tag{5-30}$$

The energy operators are the non-equilibrium solvent Hamiltonians, e.g.,

$$\mathsf{H}_{SP} = \mathsf{H}_S(q - q_P^0) + \mathsf{J}_P$$

with the eigenvalue equation

$$H_{SP}\Phi^0_{Pn'}(q) = E^0_{Pn'}\Phi^0_{Pn'}(q)$$

The operator J_P has the eigenvalue J_P which is effectively the energy of the interaction of the ions and the transferring electron plus the complete equilibrium solvation free energy of the products at the interreactant separation R. Similarly, for the reactants there are H_{SR}, J_R, and J_R.

Equation 5–30 has been integrated for several cases. One interesting case is that for low temperatures, where the theory suggests that quantum effects may occur. The original paper[13] and the discussion of Marcus[19] should be consulted for details. The high-temperature limit is defined by the condition $kT \gg \hbar\omega_0$ for the case of a single solvent frequency. For the case of a spectrum of solvent frequencies, an upper frequency ω_D is established in the same manner used by Debye in his treatment of heat capacities of solids. In the latter case, the high-temperature limit is $kT \gg \hbar\omega_D$. In either case the limit is restrictive, since at room temperature it requires $(\omega_0/2\pi) \ll 10^{13}$ sec.$^{-1}$ or $(\omega_D/2\pi) \ll 10^{-13}$ sec.$^{-1}$ which is even more restrictive. It is interesting that the dielectric relaxation time of water is about 10^{-12} sec. so that the lifetime of the activated complex, determined in this model by solvent dipole reorganization, cannot be less than this and might be somewhat larger. Hence the high-temperature limit of Levich and Dogonadze is not strictly appropriate.

Using the formulation of Marcus,[19] which is convenient for comparison with results presented in Chapter 6, the high-temperature value of k_{RP} according to the single-frequency solvent vibration case of Levich and Dogonadze is given by

$$k_{RP}(R) = \sqrt{\frac{\pi}{kT\lambda_0}}\,\frac{|L_{RP}|^2}{h}\,\exp\left[(J_P - J_R - \lambda_0)^2/4\lambda_0 kT\right] \qquad (5\text{--}31)$$

with

$$\lambda_0 = \frac{c(\Delta e)^2}{2}\left(\frac{1}{2R_a} + \frac{1}{2R_b} - \frac{1}{R}\right)$$

where

$L_{RP} = H'_{RP}$ of the previous discussion.
$\Delta e =$ charge transferred.
$R_a =$ radius of ion a.
$R_b =$ radius of ion b.
$(J_P - J_R) =$ standard free energy of the reaction (equilibrium solvation) at the separation R (zero for a homonuclear electron-transfer reaction).

Equation 5–31 is the most useful form of k_{RP}, but its derivation involves a number of important steps which are most easily understood after reading the treatment of the non-equilibrium polarization problem given by Marcus.[6,9]

The rate constant given in Eq. 5–31 is based on standard states of reactants or products, at fixed R. The electrical work required to bring the reactants to this distance from infinite separation must be taken into consideration, and the contributions from activated complexes with different values of R should be summed over R with proper weighting factors. Both operations are performed in

$$k_{RP} = 4\pi \int_0^\infty \kappa_{RP}(R) e^{-\omega(R)/kT} R^2 \, dR \qquad (5\text{–}32)$$

$$k_{RP} = 4\pi \bar{R}^2 \kappa_{RP}(\bar{R}) e^{-\omega(\bar{R})} \int_0^\infty \delta R \qquad (5\text{–}33)$$

The electrostatic work factor $\omega(R)$ is $Z_a Z_b e^2 / \varepsilon R$ with Z_i the charge on ion i and ε the macroscopic dielectric constant of the solvent. The factor $4\pi \bar{R}^2$ in Eq. 5–33 is due to implicit integration over angles as well as R. This integration corrects for the configurational entropy of the activated complex. It is not apparent that the lost entropy of mixing is accounted for in this formulation. If $\bar{R}$ is the value of R maximizing the integrand, k_{RP} is adequately approximated by Eq. 5–33 and $\int_0^\infty \delta R = \delta R$ is the narrow range of values of R in which the integrand of Eq. 5–32 is significant. The reactants are assumed spherical in Eq. 5–32.

A comparison of Eq. 5–31 with the results of Marcus and of Hush, discussed in Chapter 6, shows that this case of Levich and Dogonadze, which most closely corresponds to the case treated by Marcus and Hush, leads to an identical formulation as regards solvation of the reactant ions and solvent rearrangement in the formation of the activated complex. The methods are quite different but the results at the level of useful approximation are the same. In Eq. 5–31, k_{RP} is roughly equivalent to the fusion of the Zener-Landau-Steuckelberg treatment of the transmission coefficient with the Marcus-Hush result for solvent reorganization. The coefficients are different and simpler, but the numerical results are not much different from those obtained using the Z.L.S. formulation. More information is required in the latter so that Eq. 5–31 is an approximation to the latter and thus not necessarily an improvement except in ease of calculation. In any case, we are still uncertain as to the source of H'_{RP} and its value. Levich and Dogonadze have examined the two solvent-electron interactions which form $L_{rr'}$ (Eq. 5–21) though with no conclusive results. We shall postpone further comparisons of this theory with that of Marcus and Hush until Chapter 6. Dogonadze[167,168] has modified the theory along classical lines for application to adiabatic reactions, but the results are very little different from those of Marcus and Hush for this type of reaction and need not be discussed.

Electron Transfer Through Bridging Ligands in Inner-Sphere Reactions

Although there have been no significant attempts to examine quantitatively the expected magnitude of the transmission coefficient for bridged electron-transfer reactions in general, Halpern and Orgel[12] made some interesting, though quite approximate calculations of this coefficient for electron transfer through long bridging ligands on the assumption that these reactions are non-adiabatic, an assumption which, as mentioned at the beginning of this chapter, may not be correct. Specifically their treatment applies to bridged inner-sphere reactions, but it can be modified to include transfer in outer-sphere reactions through "bridges" formed in a casual contact of the free ends of two conducting ligands in the inner shells of the reactants, each ligand being in the first coordination sphere of one metal ion only.

As we have seen, the rate constant for

$$X_R^\ddagger \xrightarrow{k_{RP}} X_P^\ddagger \tag{5-34}$$

in which the activated complex goes from the electron configuration of the reactants' activated complex to that of the products' activated complex, is given by

$$k_{RP} = \frac{c_P^* c_P}{\tau} = \frac{(H'_{RP})^2}{\hbar^2} \tau \tag{5-35}$$

and the rate of reaction (5-34) is equal to $k_{RP}[X_R^\ddagger]$. As pointed out by Halpern and Orgel,[12] the macroscopic rate is linear in the reaction time t and k_{RP} is independent of t as it should be. The time τ appearing in Eq. 5-35 is the average lifetime of an activated complex, i.e., the length of time a favorable nuclear configuration exists, and is independent of H'_{RP} and of reaction time, t.

Equation 5-35 can be written

$$k_{RP} = \pi^2 \nu^2 \tau \tag{5-36}$$

since ν, the frequency of interconversion of reactant and product electron configurations, is approximately given by

$$\nu = \frac{2|H'_{RP}|}{h} \tag{5-37}$$

As an example of the type of expression obtained for ν, we will consider electron transfer between A and B^+ in a symmetrical, bridged activated complex. A and B^+ are identical except for an extra electron on A in $X_R^\ddagger$. The bridge, X^-, contains a closed shell of two electrons in a molecular orbital (MO) of proper symmetry to combine with the A and B^+ MO's. The molecular wave function for the reactant electron configuration (ϕ and $\bar{\phi}$ refer to electrons with spin α and β, respectively) is, neglecting the Franck-Condon factors which we shall assume constant for all cases compared, given by

$$\Psi_R = \frac{1}{\sqrt{3!}} |\phi_A \phi_X \bar{\phi}_X| \tag{5-38}$$

in which ϕ_i is the MO of the indicated species and the overlap integral S_{AX} has been neglected compared to unity because it must be very small if the weak interaction approximation is to be valid. Similarly, the molecular wave function for the product electron configuration is

$$\Psi_P = \frac{1}{\sqrt{3!}} |\phi_B \phi_X \bar{\phi}_X| \tag{5-39}$$

The first-order energies are given by

$$E_\pm = \frac{E_R \pm H_{RP}}{1 \pm S_{RP}} \tag{5-40}$$

and their difference (since ν is given more accurately by $\nu = \Delta E / h$) by

$$\Delta E = E_- - E_+ = 2 \, |-H_{RP} + S_{RP} E_R| \tag{5-41}$$

where S_{RP}^2 in the denominator has been neglected compared to unity and where the complete Hamiltonian has been retained.

The overlap integral $S_{RP} = \langle \Psi_R | \Psi_P \rangle$ is given by

$$S_{RP} = S_{AB} - S_{AX} S_{BX} \tag{5-42}$$

Introduction of Eq. 5–42 into Eq. 5–41 yields

$$\Delta E = 2 \, |-H_{RP} + (S_{AB} - S_{AX} S_{BX}) E_R| \tag{5-43}$$

If the overlap integrals were set equal to zero, Eq. 5–43 would give Eq. 5–37 for the frequency because then $H_{RP} \cong H'_{RP}$.

Introduction of Eqs. 5–38 and 5–39 into H_{RP} gives

$$H_{RP} = \langle \phi_A(1) \phi_X(2) \bar{\phi}_X(3) \, |H| \, \phi_B(1) \phi_X(2) \bar{\phi}_X(3) \rangle$$
$$- \langle \phi_A(2) \phi_X(1) \bar{\phi}_X(3) \, |H| \, \phi_B(1) \phi_X(2) \bar{\phi}_X(3) \rangle \tag{5-44}$$

As pointed out by Halpern and Orgel, the first term in Eq. 5–44 may be loosely identified with "direct exchange" between A and B and the second term with "double exchange," i.e., a concerted transfer of an electron from A to X and from X to B. If low-lying excited electron configurations are mixed with ground-state electron configurations, "superexchange" terms result. Examples of excited configurations which may contribute are $A^+X^{-2}B^+$ and AXB. The process may be a virtual one involving an unoccupied orbital of the ligand, as for $A^+X^{2-}B^+$, or the vacating of an occupied orbital of the bridge, as for AXB. A combination of these processes leading to net excitation of the bridge is possible if there is a large over-all negative free-energy change. Introduction of contributions from these excited configurations to Ψ_R and Ψ_P means that higher-energy electron configurations (in the approximation employed) contribute to the total electronic interaction in the activated complex; it does not mean that these states are formed in intermediates. When they are formed as relatively long-lived intermediates as, for example, in the reduction of the maleato-cobalt(III) complexes by Cr^{+2}, and their formation is a rate-determining step (see page 66), the product wave function for the step is that of the intermediate and not that of the ultimate products. In such cases the bridging molecule can react as an ion. Thus *cis-trans* isomerization and ester hydrolysis of the bridge has been observed. If the intermediate ion form of the bridge is metastable, the electron-transfer process has been called a "chemical mechanism."

The terms which are loosely identified as direct, double, and super-exchange owe their existence to the approximate form of the molecular wave function used. If the energy splitting, ΔE, which determines ν, were calculated with exact eigenfunctions for the upper and lower energy surfaces (surfaces *2* and *1*, respectively, in Fig. 5–1), these separate terms would not appear.

When the bridging group contains a conjugated double-bond system, Halpern and Orgel were able to show that ν is proportional to the mobile bond order[169] between the atoms of the bridge to which A and B^+ are attached.[12] The mobile bond order (MBO) is defined by

$$\text{MBO} = \sum_{\text{all } \pi \text{ electrons}} a_{ij} a_{kj} \qquad (5\text{–}45)$$

The constants a_{ij} are the coefficients of the jth atomic orbital in the ith MO. The mobile bond order has been proposed as a measure of conjugation and, as applied here, is a measure of electron "conduction" via the relevant pathway in the bridge. In other words, we may associate with a given path of conjugation a conductance property which is, in turn, an important source of information about the electronic wave functions of the bridge. Hence electron-transfer experiments may provide such information. Saturated bridging ligands have poor conductance. Even bridges with high conductance may be inefficient in electron transfer if the symmetries of the conduction MO and A and B^+ do not match. In the $Co(III) + Cr^{+2}$ reactions involving conjugated bond systems between the metal ions, the bridge π MO's and the Cr(II) MO occupied by the transferring electron are not of correct symmetry, in the "perfect octahedron" approximation, to combine in the formation of an extended $Cr(II)\text{-}X^-$ MO. Assumption of a perfect octahedron of identical ligands for the spin-free Cr(II) complex ion results in the placing of the last electron in an orbital of e_g symmetry. This Cr(II) MO is of incorrect symmetry to combine with the π MO's of the conjugated double-bond system. The transfer of an electron from Cr(II) to the bridge would be symmetry-forbidden. Likewise, if the transferring electron had to transfer into an AO of e_g symmetry in a spin-paired Co(II) complex, the transfer of the electron from the π MO's of the bridge to the cobalt e_g AO would be symmetry-forbidden. However, distortions due to electronic degeneracy (Jahn-Teller effect), dissimilar ligands, and asymmetric vibrations will mix the e_g MO's with MO's of symmetry which can combine with the bridge MO's.

Halpern and Orgel have calculated the probability for direct exchange and double exchange, using simple estimates of the mobile bond order. In addition to finding poor conductance for saturated ligands, which, however, is not so poor that it can be ignored for short conductance pathways in the bridge, they find a general tendency for bond order to fall off with increasing length of conjugated path as well as a small alternation effect in which the bond order between atoms separated by an odd number of atoms is zero. The conductance problem is not limited to bridged activated complexes. For example, Taube and co-workers have found that ligands with unshared pairs of electrons function much more efficiently as conductance pathways in outer-sphere electron-transfer reactions than do ligands without such electron pairs. Thus if only one ligand of a reactant has an unshared pair, electron transfer may occur through this ligand alone. If none of the ligands have such pairs, it appears probable that the preferred mechanism is through an inner-sphere activated complex. This is an interesting area of study, arising fron Taube's finding that bridging orbitals can function as conduction pathways. Further work of this type may also give important information about the importance of non-adiabatic electron-transfer reactions.

6
Adiabatic Electron Transfer

In Chapter 5 we found that the Ehrenfest distinction between adiabatic and non-adiabatic processes was not an absolute one except in the trivial limit of zero H'_{RP}. It was thus necessary to introduce the "degree of adiabaticity" as a relative concept dependent on some appropriate characteristic time for the process under consideration. For electron-transfer reactions this time is the lifetime of the activated complex, and its effect is to divide the possible values of H'_{RP} into two classes. The very small values such that the electron-transfer probability during the lifetime is considerably less than unity were said to define the group of non-adiabatic processes. All electron-transfer processes with values of H'_{RP} sufficiently large to effect unit probability of transfer during the lifetime then automatically fell into the adiabatic-reaction group. It is the latter group to which we now turn our attention.

Reactions of the adiabatic group lie between two limits. The first limit is established when H'_{RP} is sufficient to establish unity probability but its contribution to the energy of the activated complex is negligible. In this case, it is reasonable to recognize degenerate pairs of reactants' and products' activated complexes as was done throughout Chapter 5. The electronic description of the two complexes in each pair is different, and electron transfer occurs in an essentially discontinuous process as the system undergoes electronic rearrangement from one state to the other. This kind of process might have been called non-adiabatic since the wave functions never depart very much from the zero-order functions, but to do so would be to throw away the utility of the terms for this discussion.

The other limit for adiabatic processes is that in which H'_{RP} is sufficiently large to make a significant contribution to the energy and in this limit the electrons and nuclei are always in equilibrium. There is a single electronic state of the activated complex, the lower-lying of the true electronic states for the strongly interacting complex, and it is no longer possible to recognize reactants' and products' activated complexes except in the sense of canonical structures in a system having appreciable resonance energy. The activated complex now has the properties of a single molecule and its wave functions cannot be adequately estimated in terms of the wave functions of the two collision complexes. All adiabatic electron-transfer reactions lie between these two limits but, as discussed in Chapter 1, we can do little with reactions involving large resonance energy, i.e., strong-interaction activated complexes, and must thus confine ourselves to those with weak-interaction activated complexes. In doing so we shall have to restrict our discussion to the theories of R. A. Marcus and N. Hush; since these theories lead to nearly identical rate expressions, they ought to be known simply as the Marcus-Hush theory, though Marcus has carried the development farther than Hush and his more recent papers cover the subject in considerably greater detail. It should also be noted that Marcus has provided alternative approaches in some stages of the development which considerably illuminate the fine points of electron-transfer mechanisms. The reader will do well to examine the original papers of both Marcus and Hush, since there is insufficient room to do them justice here. Indeed, the formulation we will give is in general simpler and more practical than that of either author but correspondingly less rigorous and less illuminating. We shall, however, compare the developments of the two authors in some detail, since their respective models represent the two limits for adiabatic electron-transfer reactions involving weak interaction in the activated complex. The difference will provide an interesting point of departure for this discussion.

Marcus assumes that H'_{RP}, though sufficiently large to guarantee a transmission probability of unity, is still so small that the electron is confined to the original reductant before its migration and to the reduced product after migration. In other words, he recognizes a reactants' and a products' activated complex, and the condition of equal free energy of these complexes determines the ligand and solvent rearrangements necessary to produce the activated complexes. These nuclear rearrangements are not, however, in equilibrium with the electron distribution of either reactants' or products' activated complex but rather with a hypothetical equilibrium distribution which places part of the migrating electron on one reactant and part on the other. This distribution he calls the *equivalent equilibrium distribution* and the fraction of hypothetical charge transferred enters as a major parameter in the theory. Marcus has examined in some detail the properties of the equivalent equilibrium electron distribution to show that the concept is valid and that rigorous deductions about electron-transfer reactions can be made with it. His theory is built on this distribution.

Hush implicitly assumes that H'_{RP} is large enough to allow the electron to move freely from one reactant to the other in the activated complex, and a major parameter of his theory is also the fraction of charge transferred between reactants. However, the distribution is not a hypothetical one and the ligands and solvent molecules are considered to be in true equilibrium with the electronic distribution at all times. There is a single activated complex with the migrating electron distributed over both reactants.

Although the model adopted by Marcus is generally more realistic for weak-interaction mechanisms of electron transfer, it is also true that as H'_{RP} becomes progressively larger the model of Hush becomes increasingly appropriate and that of Marcus less appropriate. There is no sharp dividing line, so the two theories must give identical results at least in some range of H'_{RP}. Marcus has, in fact, been able to show that Hush's results are equivalent to his own for certain cases. Hence from the practical point of view, the two models are equivalent.

INTRODUCTION OF TIME INTO THE RATE-CONSTANT EXPRESSION

The method used to introduce time into the rate-constant expression for non-adiabatic electron-transfer reactions depends on the mixing perturbation assumed effective in controlling the magnitude of the transmission coefficient. It is by no means clear what choice should be made in any given case. If the perturbation is associated with a ligand and solvent rearrangement coordinate, the conventional factor, kT/h, of absolute rate theory could be used. In the treatments discussed in Chapter 5 this association is not explicitly assumed, though it might be for special cases. Instead, a frequency of application of the perturbation replaces the factor kT/h so that each choice of perturbation is properly associated with its own transmission coefficient and frequency of application. The free-energy change associated with formation of the activated complex is then a true standard free-energy change and not the usual "free energy of activation." The procedure is consistent but does not help much in practical applications of the theory. The practical situation is much improved for outer-sphere adiabatic electron-transfer reactions since, for these, time must enter through a collision number for reactants. Collisions in solution are complicated by cage effects, so that one must distinguish between encounter numbers and effective collision numbers, but the problem is one about which much is known empirically and theoretically. Quite good theories exist for such calculations, and there is a variety of empirical information, particularly from efficient quenching reactions of excited states, which will generally support the calculation of a very good estimate of the effective collision number. This fortunate situation makes the theoretical study of adiabatic outer-sphere reactions particularly attractive. Attempts to provide theoretical treatments for adiabatic electron-transfer reactions have been confined to this kind of reaction so that this chapter, which provides condensed versions of these treatments, is largely devoted to such reactions.

For inner-sphere activated complexes in electron-transfer reactions, complications are introduced by the two reactants sharing one or more inner-shell ligands. Large energies may be involved in forming these complexes, and they will be difficult to estimate. Of course, it is also possible that the over-all rate can be limited by formation or destruction of these complexes rather than the electron-transfer step with its associated ligand and solvent rearrangements. When such reactions are adiabatic, conventional rate theory must be used, with time appearing in the factor kT/h. These reactions are nearly as difficult to treat quantitatively as conventional chemical reactions involving strong-interaction activated complexes, so there is little we can add in this chapter to simplify the general problem.

For adiabatic electron-transfer reactions of the outer-sphere type, the effective collision number Z and the standard free-energy change in forming the activated complex from the *reactants' collision complex* must be estimated. These two quantities interact, and one must be consistent in making the estimates. Marcus has treated the matter in a way which takes full advantage of known information about collision numbers in solution. He chooses to consider that the collision complex is formed from uncharged reactants. It is then necessary to include a coulombic free-energy term for this complex in his free-energy expression. We shall now discuss these estimates of the Z and the free-energy change in detail.

The parallelism between solution binary collision processes which lead to de-excitation from higher electronic states and outer-sphere electron-transfer processes is very close, so close in fact as to be identical when electron-transfer mechanisms are responsible for de-excitation. When the quenching processes are effective on each encounter, the measured collision numbers provide an entirely suitable basis for estimating collision numbers in adiabatic reactions; this is true whether the quenching process be of the electron-transfer type, of the type in which the quencher deactivates by exchange of spin momentum and energy as, for example, when oxygen quenches triplet states by itself becoming excited to the lowest $^1\Delta$ state, or of the type in which the quencher simply provides a magnetic perturbation. We mention these mechanisms of quenching simply to point out that the theory of Marcus or the very similar theory of Hush could be easily and usefully extended to quenching processes of all types.

Collision numbers vary with size of reactants, with the characteristics of the solvent, especially viscosity, and with the coulombic interactions. Ignoring the last factor, which goes into the effective free energy of activation, there are still several complications, which must be given specific consideration in using Z values from quenching experiments, in estimating Z values for adiabatic electron-transfer reactions. The most important are:

1. In quenching processes used as models, the quencher must be effective in each encounter.
2. There is an uncertainty in the relation between van der Waals' contact diameters and effective collision diameters in quenching, since the latter are often larger than the former.

3. There may be uncertainty as to changes in degrees of freedom in forming the collision complex. If any specific orientation of the reactant is required for efficient quenching or efficient electron transfer, there will be fewer rotational and more vibrational degrees of freedom in the collision complex than when such steric requirements are lacking.
4. There may be weak bonding between the reactants in the collision complex. Hydrogen or hydrophobic bonding can stabilize the complex and also change the distribution of degrees of freedom.

Thus far, little attention has been paid to the estimation of Z for adiabatic electron-transfer reactions since, in general, the errors introduced in such estimates are small relative to those introduced in estimating the free-energy of activation. Marcus[9] has discussed the problem and chooses to use a single universal value, for water at room temperature, of 2.5×10^{11} M^{-1} sec.$^{-1}$. At present, the use of a single Z value is reasonable, but quenching experiments suggest that Marcus' estimate is perhaps as much as tenfold too large. There is considerable literature on the subject; the reader wishing to make a better estimate for any specific case is referred, in particular, to the work of Collins and Kimball,[170] Weller,[171] and Noyes.[172]

THE TOTAL FREE ENERGY OF ACTIVATION

Following the procedure of Marcus, we calculate the total free energy of activation as the free energy of forming the activated complex from the uncharged collision complex. To carry out a similar calculation for a non-adiabatic outer-sphere electron-transfer reaction using the formalisms of Chapter 5, we should then simply have to add the free energy of forming the collision complex from the separated reactants. We point out again, however, that there is an important subtle difference between this calculation and that of conventional rate theory, quite apart from the method of including time. In absolute rate theory, the activated complex is an ordinary member of the allowed states of the system, and its concentration is the true equilibrium concentration under the experimental conditions. In the theory of Marcus, the activated complex is not a true equilibrium state but is a hypothetical complex having the equivalent equilibrium electron distribution, with ligands, solvent, and solvent ions in equilibrium with this distribution. Marcus has shown that the free energy of formation of this hypothetical state from the collision complex does provide the correct estimate of the probability of occurrence of the activated complex and thus gives the desired exponential term in the total rate-constant expression; in Hush's treatment, the activated complex is of the conventional type. In using the Marcus approach in strictly correct fashion, all rearrangements should take into account the equivalent equilibrium distribution. Thus the effective force constants for ligand or solvent rearrangement should be those characteristic of this distribution if the energy and entropy balance conditions are to be satisfied exactly by the activated complex. This is a rather elaborate procedure and, at the present time, often an irrelevant one in view of the limited accuracy of the treatments, since the corrections do not contribute large errors. For this reason, in the following we may occasionally approximate the strictly correct calculation both to allow brevity of exposition and to provide sufficiently simple expressions for practical application.

According to the method of Marcus, the free energy of activation is given by

$$\Delta F^* = \Delta F_c^{\ddagger} + \Delta F_i^{\ddagger} + \Delta F_0^{\ddagger} + \Delta F_e^{\ddagger} \tag{6-1}$$

in which $\Delta F_c^{\ddagger}$ is the free-energy change due to coulombic interaction of reactants at the nuclear configuration of the collision complex, $\Delta F_0^{\ddagger}$ is the rearrangement free energy of the outer-sphere solvent and ion atmosphere in the field of the equivalent equilibrium electron distribution of the activated complex, $\Delta F_i^{\ddagger}$ is the rearrangement free energy of the inner-sphere ligands in the field of the equivalent equilibrium electron distribution or of skeletal distortions of molecules other than complex ions, and $\Delta F_e^{\ddagger}$ is an entropic quantity due to change of electronic quantum numbers when electron transfer occurs. Promotional and vibronic distortion changes of electronic quantum numbers can be included in this term.*

We now see that ΔF^* is related to the conventional free energy of activation $\Delta F^{\ddagger}$ by

$$\Delta F^{\ddagger} = \Delta F^* + RT \ln (kT/Zh) \tag{6-2}$$

THE CHANGE-OF-MULTIPLICITY CONTRIBUTION

The $\Delta F_e^{\ddagger}$ contribution can never be very large, but its inclusion emphasizes the possible importance of electronic quantum-number changes in forming the activated complex when electron migration involving the electronic ground states is significantly restricted by selection rules. If change of electronic quantum number, usually multiplicity, occurs on formation of the activated complex, $\Delta F_e^{\ddagger}$ is given by

$$\Delta F_e^{\ddagger} = RT \ln (W^{\ddagger}/W_R) \tag{6-3}$$

in which $W^{\ddagger}$ is the multiplicity of the activated complex and W_R that of the ground-state reactants. If such changes occur on electron migration in the activated complex, $\Delta F_e^{\ddagger}$ is given by

$$\Delta F_e^{\ddagger} = RT \ln (W_R^{\ddagger}/W_P^{\ddagger}) \tag{6-4}$$

in which $W_R^{\ddagger}$ and $W_P^{\ddagger}$ are the multiplicities of the reactants' and products' activated complexes, respectively. In Marcus' formulation, $W_R^{\ddagger}$ and $W_P^{\ddagger}$

---

* In addition to the terms given for ΔF^* there should be terms for contributions from hydrogen or other types of bonding between reactants. These have been neglected to date.

need not be equal; in Hush's formulation, the two must be equal. The problem has not yet been examined in detail. It has considerable intrinsic interest despite the fact that $\Delta F_e^{\ddagger}$ is small.

COULOMBIC FREE ENERGY

The coulombic force between the two complex ion reactants is repulsive when the charges on the complex ions have the same sign. The contribution to ΔF^* is then positive, otherwise negative; it is approximated, using Debye-Hückel theory, by[90,173]

$$\Delta F_c^{\ddagger} = \frac{z_1 z_2 e^2}{\varepsilon r_{\ddagger}} \exp\left[-r_{\ddagger}\left(\frac{8\pi e^2 N_0 \mu}{1000\varepsilon kT}\right)^{1/2}\right] \tag{6-5}$$

where z_1 and z_2 are the valences of reactants 1 and 2, $r_{\ddagger}$ is the distance between centers of the two reactants, μ is the ionic strength, and the other quantities have been defined. The distance $r_{\ddagger}$ in outer-sphere activated complexes is usually taken as the sum of the radii of the two reactants *after* inner-coordination-sphere rearrangement. At 25°C with $r_{\ddagger}$ in angstroms, Eq. 6-5 reduces to

$$\Delta F_c^{\ddagger} = \frac{4.22 z_1 z_2 \times 10^{-0.143 r_{\ddagger}\sqrt{\mu}}}{r_{\ddagger}} \text{ kcal./mole} \tag{6-6}$$

The accompanying entropy change is

$$\Delta S_c^{\ddagger} = \frac{z_1 z_2 e^2 e^{-\gamma}}{\varepsilon^2 r_{\ddagger}}\left(\frac{\partial \varepsilon}{\partial T}\right) + \frac{z_1 z_2 e^2 \gamma e^{-\gamma}}{2\varepsilon r_{\ddagger}}\left(\frac{1}{T} + \frac{\partial \ln \varepsilon}{\partial T}\right) \tag{6-7}$$

where γ is the exponent in Eq. 6-5. For aqueous solutions, $\varepsilon = 78.5$, and $(\partial \varepsilon / \partial T) = -0.36$ deg.$^{-1}$ at 25°C, so that Eq. 6-7 reduces to

$$\Delta S_c^{\ddagger} = -z_1 z_2 \times 10^{-0.143 r_{\ddagger}\sqrt{\mu}}\left(\frac{19.3}{r_{\ddagger}} + 0.83\sqrt{\mu}\right)\frac{\text{cal.}}{\text{deg. mole}} \tag{6-8}$$

where $r_{\ddagger}$ is now in angstroms. For typical values of $\mu = 1$ and $r_{\ddagger} = 7$ Å, $\Delta S_c^{\ddagger} = -0.36 z_1 z_2$ cal./deg. mole. Thus it is seen that the exponential appearance of $\sqrt{\mu}$ greatly reduces the contributions that $\Delta F_c^{\ddagger}$ and $\Delta S_c^{\ddagger}$ make to the complete activation parameters; they could be neglected without introducing significant error when the ionic strength is high.

Hush[15] has derived an expression for the coulombic potential energy in terms of a parameter $\lambda^{\ddagger}$, which is the fractional charge transferred to the oxidant in the activated complex and which is closely identified with $-m$, a parameter of Marcus' referred to in the next section. However, this equation will not be given, because it contains a contribution from the outer coordination spheres of the reactants, and it is preferred here to consider the various contributions separately for reasons of clarity and brevity.

INNER-SPHERE REARRANGEMENT FREE ENERGY

For any given reaction, special information is required about changes in ligand geometry required for mixing of electronic states in such a way as to satisfy Franck-Condon restrictions. We shall consider only the simplest kind of ligand reorganization, such as changes in vibrational coordinates associated with simple breathing motions of the ligands and required to produce the degenerate activated complex.

The free-energy change of the inner-coordination-sphere rearrangement can be expressed in terms of vibrational partition functions by

$$\Delta F_i^{\ddagger} = -RT \ln \left(\frac{Q_{\ddagger,v}}{Q_{A,v}Q_{B,v}} \right) + \Delta U_i^{\ddagger} \qquad (6\text{-}9)$$

in which $\Delta U_i^{\ddagger}$ is the difference in zero-point energies between $X^{\ddagger}$ and the reactants A and B and the Q's are the vibrational partition functions of the indicated species. This is the expression obtained by Marcus.[9] It is assumed that the vibrational energy is separable from the rotational, translational, electronic, and outer-coordination-sphere energies. The vibrational partition functions are probably approximately equal to unity, so that Sutin[18] has taken

$$\Delta F_i^{\ddagger} = \Delta U_i^{\ddagger} \qquad (6\text{-}10)$$

For a reaction such as

$$M(H_2O)_n^{z-1} + {}^*M(H_2O)_n^{z} \rightleftharpoons M(H_2O)_n^{z} + {}^*M(H_2O)_n^{z-1}$$

the work required to give a common value $r_{\ddagger}$ to the M—OH$_2$ distance in each reactant has been given by Sutin[18] in a simplified treatment as

$$\Delta F_i^{\ddagger} = \frac{nf_1}{2}(r_1 - r_{\ddagger})^2 + \frac{nf_2}{2}(r_2 - r_{\ddagger})^2$$

in which f_1 and f_2 are the force constants of the M—OH$_2$ bonds of the two reactants, and r_1 and r_2 are their equilibrium distances. When $\Delta F_i^{\ddagger}$ is minimized with respect to $r_{\ddagger}$ and the resulting expression rearranged, it is found that

$$r_{\ddagger} = \frac{f_1 r_1 + f_2 r_2}{f_1 + f_2} \qquad (6\text{-}11)$$

Substitution of this expression for $r_{\ddagger}$ in the equation for $\Delta F_i^{\ddagger}$ gives

$$\Delta F_i^{\ddagger} = \frac{nf_1 f_2 (r_1 - r_2)^2}{2(f_1 + f_2)} \qquad (6\text{-}12)$$

For the more general case, Marcus[10] gives

$$\Delta F_i^{\ddagger} = m^2 \lambda_i \tag{6-13}$$

$$m = -\frac{1}{2} - \frac{\Delta F^{\circ} + (\Delta F_{C,P}^{\ddagger} - \Delta F_{C,R}^{\ddagger})}{2\lambda} \tag{6-14}$$

$$\lambda = \lambda_0 + \lambda_i \tag{6-15}$$

$$\lambda_i = \sum_j \frac{f_j f_j^P}{(f_j + f_j^P)} (\Delta r_j^0)^2 \tag{6-16}$$

In these equations, ΔF° is the over-all free-energy change in the medium used for the elementary reaction step in which electron transfer occurs (not usually the standard free-energy change of the step), $(\Delta F_{C,P}^{\ddagger} - \Delta F_{C,R}^{\ddagger})$ is the difference between Eq. 6–5 for the products and for the reactants, $m^2\lambda_0$ is the rearrangement free energy of the outer coordination spheres (discussed in the next section), $m^2\lambda_i$ is the rearrangement free energy of the inner coordination spheres, the f_j and f_j^P are the force constants of the jth vibrational coordinate in a species when the species participates as a reactant and as a product, respectively, and the Δr_j^0 are the changes in bond lengths and bond angles in the reactants.

For the isotope exchange reaction between hexaquo iron(II) and iron(III) ions, Sutin[18] used f_1 and f_2 equal to 1.49×10^5 and 4.16×10^5 dyne/cm., respectively, $r_1 = 2.21$ Å, and $r_2 = 2.05$ Å to obtain $r_{\ddagger} = 2.09$ Å. Although the two ions have identical inner coordination spheres, it is seen that $r_{\ddagger}$ is not halfway between r_1 and r_2. This suggests that it is easier to compress the Fe^{+2}—OH_2 bonds than to stretch the Fe^{+3}—OH_2 bonds so that the minimum $\Delta F_i^{\ddagger}$ value is obtained at $r_{\ddagger}$ values nearer the Fe^{+3}—OH_2 bond length. The corresponding $\Delta F_i^{\ddagger}$ value is 12 kcal./mole.

The force constants f_1 and f_2 for $Fe(H_2O)_6^{+2}$ and $Fe(H_2O)_6^{+3}$, respectively, were calculated by Sutin from Eq. 2–15 for U by setting the second derivative of U with respect to r at the equilibrium distance equal to f. Only the symmetrical breathing mode of the complex ion was considered. The corresponding frequencies, equal to $\frac{1}{2}\pi\sqrt{f/\mu}$, where μ is the reduced mass, calculated by Sutin were 430 cm.$^{-1}$ and 7.8 cm.$^{-1}$. The experimental value of the frequency of this symmetrical breathing mode for $Zn(H_2O)_6^{+2}$ is 394 cm.$^{-1}$ (see Ref. 174). There is, therefore, reasonable agreement between the experimental value and the value calculated from a simple electrostatic, harmonic oscillator model.

It is, of course, not necessary to calculate force constants and then to use these force constants to calculate the free energy of inner-sphere ligand displacements according to the harmonic (or other) oscillator model. If all that is required is $\Delta F_i^{\ddagger}$, then this quantity can be calculated directly from the potential-energy functions (U in this approximation) for the two reactants.

The $\Delta F_i^{\ddagger}$ value calculated directly from the U's will differ from the $\Delta F_i^{\ddagger}$ value calculated from force constants and the harmonic oscillator model in general. The reason for this is that the U versus r curves are distinctly those for anharmonic oscillators, U increasing more rapidly at $r < r_e$ than at $r > r_e$. The fact that the U versus r curve is steeper at $r < r_e$ than it is at $r > r_e$ means, for example, that $Fe(H_2O)_6^{+2}$ will not be compressed as much as, and that $Fe(H_2O)_6^{+3}$ will be extended more than, the harmonic oscillator model shows. As a result, the common $r_{\ddagger}$ value will be nearer the middle value of 2.13 Å than it is in the harmonic oscillator model. It must be noted that $\Delta F_i^{\ddagger}$ should be calculated from expressions for U such as Eq. 2–19 in which variation of the induced moment, μ_i, with distance r has been taken into account; when $\Delta F_i^{\ddagger}$ is calculated from expressions for U such as Eq. 2–15 with a fixed value of μ_i (such as the value of μ_i at the equilibrium distance r_e), error will be introduced needlessly.

In the approximation that the inner-coordination-sphere energy is independent of the outer solvation spheres, we may calculate $\Delta F_i^{\ddagger}$ from the U functions by

$$\Delta F_i^{\ddagger} = (U_1^{\ddagger} - U_1^0) + (U_2^{\ddagger} - U_2^0) \qquad (6\text{--}17)$$

where $U_1^{\ddagger}$, $U_2^{\ddagger}$ are the potential energies of reactants 1 and 2 in a common configuration specified by $r_{\ddagger}$, and U_1^0, U_2^0 are the potential energies of 1 and 2 with their equilibrium distances. The value of $r_{\ddagger}$ which minimizes $\Delta F_i^{\ddagger}$ is obtained from

$$\frac{\partial (\Delta F_i^{\ddagger})}{\partial r_{\ddagger}} = \frac{\partial U_1^{\ddagger}}{\partial r_{\ddagger}} + \frac{\partial U_2^{\ddagger}}{\partial r_{\ddagger}} = 0 \qquad (6\text{--}18)$$

When the values of 13.9×10^{-82} and 17.0×10^{-82} erg. cm.^{9}, obtained from minimizing Eq. 2–19 with respect to r for $Fe(H_2O)_6^{+2}$ and $Fe(H_2O)_6^{+3}$, respectively, are inserted for β in Eq. 2–19, along with the values given in Chapter 2 for the other constants, it is found that a value of $r_{\ddagger}$ between 2.11 and 2.12 Å satisfies Eq. 6–18. Of course, Eqs. 6–17 and 6–18 can be applied to ions of symmetry other than octahedral if the proper expression for U is derived.

Sacher and Laidler[16] did not consider the inner-coordination-sphere energy to be independent of the outer-sphere solvation energy. Instead, they assumed a concomitant increase and decrease of solvation energy as the inner coordination sphere decreased or increased in size. The total change of inner- and outer-sphere rearrangement enthalpy was calculated from a plot of an expression similar to Eq. 2–27 for each ion in the hexaquo Fe(II) + Fe(III) exchange. The terms in Eq. 2–27 which represent change of solvation energy as the radius of the complex ion changes are calculated from the Born expressions in Eqs. 2–12, 2–13, and 2–14. *Therefore, in the treatment of Sacher and Laidler the solvent in the outer solvation shells is in equilibrium with the charge on the ion for all radii of the complex ions.** In partic-

---

* Note that we have ignored throughout changes of force constants due to the change in electronic arrangement into the equivalent equilibrium distribution.

ular, when the $Fe(H_2O)_6^{+2}$ ion contracts to the configuration specified by $r_{\ddagger}$, the outer solvation shells will have the solvent dipole configuration characteristic of that due to the $+2$ charge on the complex ion; when the $Fe(H_2O)_6^{+3}$ ion expands to the configuration specified by $r_{\ddagger}$, the outer solvation shells will have the solvent dipole configuration characteristic of that due to the $+3$ charge on the complex ion. If, at this configuration, an electron is transferred from the $+2$ ion to the $+3$ ion, the dipole configuration characteristic of the $+2$ ion will find itself in the field of a $+3$ ion; since the dipole arrangement around a $+2$ ion will not be the same as that around a $+3$ ion, this non-equilibrium state arising from electron transfer will represent a higher energy state for the solvent dipoles. Likewise, after electron transfer the solvent, which was in equilibrium with the field of a $+3$ ion before electron transfer, around the resulting $+2$ ion will be in a non-equilibrium, high energy state. Unless the required energy can be absorbed from the surroundings during the non-radiative electron-transfer process or unless the energy increase is within the limits set by the uncertainty principle, energy will not be conserved and the process is impossible. If it is assumed that the solvation energy change accompanying change of inner sphere radius cannot be neglected, then, in order to get the final correct solvent orientation for electron transfer, the calculation should be carried out as follows:

1. The enthalpy and free-energy changes arising from rearrangement of an inner coordination sphere to a radius $r_{\ddagger}$ and the accompanying solvation enthalpy and free-energy changes are calculated from Eqs. 2–27 and 2–28.
2. The inner coordination spheres of both reactant ions are kept constant at their respective $r_{\ddagger}$ values while the solvent is reoriented according to Marcus' original theory.[6]
3. The calculated $\Delta H^{\ddagger}$ and $\Delta F^{\ddagger}$ values are corrected for zero-point energies as in the Sacher and Laidler method.[16]

This was the method by which the free-energy changes for the inner and outer coordination spheres of the $Fe^{+2}(aq) + Fe(ph)_3^{+3}$ reaction in Chapter 3 were calculated. Marcus has provided a further improvement in this type of calculation.[11]

The expression for $\Delta F_i^{\ddagger}$ derived by Hush[15] for isotope exchange reactions in which $\Delta F° = 0$ is

$$\Delta F_i^{\ddagger} = \frac{ne\mu}{4(m-2)}\left(\frac{1}{z_1 r_1^2} + \frac{1}{z_2 r_2^2}\right) \tag{6–19}$$

where n is the number of ligands in the inner coordination spheres of reactants 1 and 2, z_1 and z_2 are the absolute values of the valences of the two ions, m is the power of r in the repulsive β term of the U functions for 1 and 2, r_1 and r_2 are the equilibrium metal-ligand distances in reactants 1 and 2, and μ is the ligand dipole moment.   In deriving Eq. 6–19, it was assumed[15] that U was given by

$$U = -\frac{n\mu ze}{r^2} + \frac{\beta}{r^m}$$

in which both β and μ were independent of charge.   The last assumption is especially crude since μ increases by approximately 1.34 debye when the charge is increased from $+2$ to $+3$ in the hexaquo iron ions.

The entropy change $\Delta S_i^{\ddagger}$ is negligible since the rather minor changes in metal-ligand bond lengths do not appreciably affect the entropy.

OUTER-SPHERE REARRANGEMENT FREE ENERGY

Marcus[6] calculated the energy required to rearrange the outer solvation spheres of the reactant ions from the equilibrium solvent orientation of the ground-state reactants in the collision complex to the non-equilibrium arrangement consistent with the equivalent equilibrium electron distribution. He considered the solvent to be a classical dielectric continuum.   The expression obtained was

$$\Delta F_0^{\ddagger} = m^2\lambda_0 \tag{6–20}$$

in which m is given by Eq. 6–14 and λ_0 is given by

$$\lambda_0 = (\Delta z)^2 e^2\left[\frac{1}{2r_1^{\ddagger}} + \frac{1}{2r_2^{\ddagger}} - \frac{1}{r_{\ddagger}}\right]\left[\frac{1}{\varepsilon_{op}} - \frac{1}{\varepsilon}\right] \tag{6–21}$$

in which $|\Delta z|$ is the valence change or the number of electrons transferred in the electron-transfer step, $r_1^{\ddagger}$ and $r_2^{\ddagger}$ are the radii of reactants 1 and 2 in the activated complex, $r_{\ddagger}$ is the distance between centers of reactants 1 and 2 in the activated complex, ε_{op} is the optical dielectric constant, i.e., the square of the refractive index, and the other quantities have been defined.   If the first coordination spheres of the reactants just touch in the activated complex, then

$$r_{\ddagger} = r_1^{\ddagger} + r_2^{\ddagger}$$

In addition, if the electron-transfer reaction is an isotope exchange reaction in which the reactants and products are identical, then

$$r_1^{\ddagger} = r_2^{\ddagger}$$

$$r_{\ddagger} = 2r_1^{\ddagger} = 2r_2^{\ddagger}$$

and, from Eq. 6–14,

$$m = -\tfrac{1}{2}$$

In water, $1/\varepsilon$ is less than 3 percent of $1/\varepsilon_{op}$ so that the expression for $\Delta F_0^{\ddagger}$ reduces to a particularly simple relation namely,

$$\Delta F_0^{\ddagger} = \frac{(\Delta z)^2 e^2}{8\varepsilon_{op}r_1^{\ddagger}} \tag{6–22}$$

It is noteworthy that $\Delta F_0^{\ddagger}$ is only slightly dependent on the value of the dielectric constant ε as long as $\varepsilon \gg \varepsilon_{\mathrm{op}}$. Therefore, if some dielectric saturation occurs outside of the first coordination spheres of the reactant ions, the effect on $\Delta F_0^{\ddagger}$ will be small and will not embarrass us. Also, $\Delta F_0^{\ddagger}$ is not particularly sensitive to the ionic radius of the metal ion because of the large additive constant, the ligand diameter, in Eq. 6–23 for r_1:

$$r_1 = r_{\mathrm{ion}} + 2r_L \tag{6–23}$$

Here r_{ion} is a crystal ionic radius of the metal ion and r_L is a radius of the ligand. For H_2O, $2r_L \simeq 2.76$ Å and a change of r_{ion} equal to approximately 0.35 Å is required to change r_1 by 10 percent. Thus $\Delta F_0^{\ddagger}$ will be nearly constant for a wide range of reactions between ions of different charge types, provided the same number of electrons is transferred in all reactions so that $(\Delta z)^2$ remains constant. $\Delta F_0^{\ddagger}$ will increase as the square of the number of electrons transferred. In Chapter 5, we observed that it was this larger increase in rearrangement free energy which was probably of most importance in reducing the probability for two-electron transfer.

For isotope exchange reactions at 25°C in water, Eq. 6–20 becomes

$$\Delta F_0^{\ddagger} = \frac{22.7(\Delta z)^2}{r_1^{\ddagger}} \frac{\mathrm{kcal.}}{\mathrm{mole}}$$

when $r_1^{\ddagger}$ is expressed in angstroms.

The expression for $\Delta S_0^{\ddagger}$ is complex because of the temperature dependence of both m and λ_0; it is

$$\Delta S_0^{\ddagger} = \frac{-m\lambda_0}{\lambda} C_1 - \frac{m\lambda_0}{\lambda^2} C_2 \left(\frac{\partial \lambda_0}{\partial T} \right) - m^2 \left(\frac{\partial \lambda_0}{\partial T} \right) \tag{6–24}$$

in which

$$C_1 = \Delta S^{\circ} + \Delta S_{c,P}^{\ddagger} - \Delta S_{c,R}^{\ddagger} \tag{6–25}$$

$$C_2 = \Delta F^{\circ} + \Delta F_{c,P}^{\ddagger} - \Delta F_{c,R}^{\ddagger} \tag{6–26}$$

$$\frac{\partial \lambda_0}{\partial T} = (\Delta z)^2 e^2 \left(\frac{1}{2r_1^{\ddagger}} + \frac{1}{2r_2^{\ddagger}} - \frac{1}{r_{\ddagger}} \right) \left(\frac{1}{\varepsilon^2} \frac{\partial \varepsilon}{\partial T} - \frac{1}{\varepsilon_{\mathrm{op}}^2} \frac{\partial \varepsilon_{\mathrm{op}}}{\partial T} \right) \tag{6–27}$$

For isotope exchange reactions, Eq. 6–24 reduces to

$$\Delta S_0^{\ddagger} = \frac{(\Delta z)^2 e^2}{8r_1^{\ddagger}} \left(\frac{1}{\varepsilon_{\mathrm{op}}^2} \frac{\partial \varepsilon_{\mathrm{op}}}{\partial T} - \frac{1}{\varepsilon^2} \frac{\partial \varepsilon}{\partial T} \right)$$

It is seen from Eqs. 6–24 and 6–25 that $\Delta S_0^{\ddagger}$ depends on ΔS°, the over-all entropy change of the reaction, with a proportionality constant less than unity. For isotope exchange reactions, $\Delta S_0^{\ddagger}$ is approximately equal to zero because the dielectric constant term is approximately $-3 \times 10^{-5}/°K$ in water.

For one-electron isotope-exchange reactions, Eq. 54 of Hush[15] for ΔF^* can be interpreted to contain a term for $\Delta F_0^{\ddagger}$, namely,

$$\Delta F_0^{\ddagger} = \lambda^{\ddagger}(1 - \lambda^{\ddagger})\left[\frac{1}{2r_1^{\ddagger}} + \frac{1}{2r_2^{\ddagger}} - \frac{1}{r_{\ddagger}}\right]\left[\frac{1}{\varepsilon_{op}} - \frac{1}{\varepsilon}\right]e^2 \qquad (6\text{--}28)$$

which is identical to Eq. 6-20, Marcus' result, since $\lambda^{\ddagger}(1 - \lambda^{\ddagger}) = m^2$ in this case. Since Hush divided the total free energy into individual contributions in a different way than did Marcus, this identification cannot be made when $\lambda^{\ddagger}(1 - \lambda^{\ddagger}) \neq m^2$.

COMPARISON OF THEORIES

The complete expression of Marcus[6,9,10] for ΔF^* is

$$\Delta F^* = \Delta F_{c,R}^{\ddagger} + m^2\lambda_0 + m^2\lambda_i \qquad (6\text{--}29)$$

where $\Delta F_{c,R}^{\ddagger}$ is given by Eq. 6-5 which, at zero ionic strength, reduces to

$$\Delta F_{c,R}^{\ddagger} = z_1z_2e^2/\varepsilon r_{\ddagger} \qquad (6\text{--}30)$$

$$m = -\frac{1}{2} - \frac{C_2}{2(\lambda_0 + \lambda_i)} \qquad (6\text{--}31)$$

$$\lambda_0 = (\Delta z)^2e^2\left[\frac{1}{2r_1^{\ddagger}} + \frac{1}{2r_2^{\ddagger}} - \frac{1}{r_{\ddagger}}\right]\left[\frac{1}{\varepsilon_{op}} - \frac{1}{\varepsilon}\right] \qquad (6\text{--}21)$$

At zero ionic strength, Eq. 6-26 for C_2 becomes

$$C_2 = \Delta F^{\circ} + \frac{(z_1 - 1)(z_2 + 1)e^2}{\varepsilon r_{\ddagger}} - \frac{z_1z_2e^2}{\varepsilon r_{\ddagger}} \qquad (6\text{--}32)$$

where $z_1 - 1$ and $z_2 + 1$ are the valences of the product ions and z_1 and z_2 are the valences of the reactant ions in a one-electron transfer reaction.

The inner-coordination-sphere rearrangement energy $m^2\lambda_i$ may be expressed in different ways, as we have seen.

The expression derived by Hush[15] for one-electron transfer reactions at zero ionic strength is

$$\Delta F^* = \frac{z_1z_2e^2}{\varepsilon r_{\ddagger}} + \lambda^{\ddagger}\left[\Delta F^{\circ} + \frac{e^2(z_1 - z_2 - 1)}{\varepsilon r_{\ddagger}}\right] + \lambda^{\ddagger}(1 - \lambda^{\ddagger})$$

$$\times \left[e^2\left(\frac{1}{2r_1^{\ddagger}} + \frac{1}{2r_2^{\ddagger}} - \frac{1}{r_{\ddagger}}\right)\left(\frac{1}{\varepsilon_{op}} - \frac{1}{\varepsilon}\right) + e\mu\left(\frac{n_1}{z_1(m_1 - 2)r_1^2} + \frac{n_2}{z_2(m_2 - 2)r_2^2}\right)\right] \qquad (6\text{--}33)$$

which is of the form

$$\Delta F^* = \zeta + \lambda^{\ddagger}\xi + \lambda^{\ddagger}(1 - \lambda^{\ddagger})\chi \qquad (6\text{--}34)$$

When Eq. 6–34 is minimized with respect to $\lambda^{\ddagger}$, it is found that $\lambda^{\ddagger}$ is given by

$$\lambda^{\ddagger} = \frac{1}{2} + \frac{\xi}{2\chi} \tag{6-35}$$

Equations 6–31 and 6–35 are very similar, and $\lambda^{\ddagger}$ is identical to $-m$ if $C_2 = \xi$ and $\chi = \lambda_0 + \lambda_i$. Comparison of ξ and C_2 shows that they are indeed equal for a one-electron transfer, since

$$\frac{(z_1 - 1)(z_2 + 1)e^2}{\varepsilon r_{\ddagger}} - \frac{z_1 z_2 e^2}{\varepsilon r_{\ddagger}} = \frac{e^2(z_1 - z_2 - 1)}{\varepsilon r_{\ddagger}}$$

but that they are unequal if more than one electron is transferred. On comparison of χ with $\lambda_0 + \lambda_i$, we find that the first term in χ is equal to λ_0 for one-electron transfer reactions only. Whether or not the second term in χ equals λ_i depends on how the inner-sphere rearrangement free energy is calculated; for one-electron isotope exchange reactions, the inner-sphere rearrangement free energy can be calculated so that λ_i is equal to the second term in χ. Thus, in this case, at least, $\lambda^{\ddagger} = -m$. In other one-electron reactions, whether these two quantities are equal or not depends on how one elects to calculate the inner-sphere rearrangement free energy in the two theories.*

It is apparent that there is little to choose between the formalisms of Marcus and Hush. Provided that one is consistent in their application, the two theories give the same results if one uses the same procedures to take care of ligand rearrangement, etc. There are some minor errors in Hush's development, and Hush has not refined his treatment to the extent to be found in the latest papers of Marcus; nor has he included so many of the minor contributions to the total rearrangement free energy. In general, the formulation of Marcus is easier to use and will probably become the dominant one in the subject. Criticisms and comparisons of the theories have been presented by both Marcus[19] and Sutin.[18] Some quantitative comparisons of theoretical predictions with experimental rate constants are given by Sutin and are presented in the section on cross reactions. Sutin also gives some predictions for activation energies and entropies, but it should be realized that existing theoretical treatments do not appear to be adequate for this separation. Indeed, as of this date there is very little understanding at experimental or theoretical levels of the temperature-dependence of electron-transfer reactions. We shall discuss this matter briefly at a later point (page 142) and now note only that it presents an outstanding area for further development of our understanding of electron-transfer reactions.

In Chapter 5, considerable space was devoted to the theory of Levich and Dogonadze since this development illustrates the problems and methods of a quantum-mechanical approach to the more general example provided by non-adiabatic electron-transfer reactions. It can also be shown that the

*If they are equal then Eqs. 6–29 and 6–33 yield identical results for ΔF^*, as may be readily shown since $\lambda^{\ddagger}C_2/\chi + \lambda^{\ddagger} - (\lambda^{\ddagger})^2 = m^2$.

final results of the theory, in the case of a single polaron frequency, no ligand reorganization, and at ordinary temperatures, are equivalent to those of Marcus. However, there are a number of omissions and simplifications in that theory and in the final results which make it less satisfactory than the most recent form of the theory of Marcus;[11] it is for this reason and the contrast with the quantum-mechanical development used by Levich and Dogonadze that we have presented the Marcus theory in Chapter 6. The adiabatic modification of the Levich and Dogonadze results provided by Dogonadze[167,168] and the theory of Hush[15] are somewhat similar but less comprehensive than the latest results of Marcus[11,19] in that they take account of fewer of the processes associated with the formation of the activated complex. There is very little difference in the results of all the theories when the same factors contributing to the free energy of formation of the activated complex are treated. The theories of Dogonadze and Hush could be brought further into line with the latest results of Marcus by considering contributions which were originally omitted.

CORRECTION FOR RESONANCE ENERGY

The calculation of theoretical ΔF^* (or $\Delta F^{\ddagger}$) values has thus far concerned those reactions in which $E_+ \simeq E_R$ (or E_P). When the energy difference $E_R - E_+$ is appreciable, the interaction energy can be calculated as in the theory of Halpern and Orgel[13] (cf. page 114) or by some more accurate method and a correction applied. The theoretical considerations of Halpern and Orgel are appropriate for outer-sphere (bridged and non-bridged) mechanisms and for bridged inner-sphere mechanisms depending on the wave functions selected to represent reactant and product electron configurations. The final expression obtained for the interaction energy H'_{RP} will depend on whether the reaction is a homonuclear or heteronuclear reaction, on the number of electrons which are considered in the interaction, on the nature and number of perturbation terms introduced in H', on the symmetries and degeneracies of the molecular orbitals (if MO theory is used) in the determinantal molecular wave function, on the extent to which overlap and multicentered integrals can be neglected, on whether configuration interaction is included, and on whether rotational and vibrational wave functions are included. In other words, the calculation of H'_{RP} is an ordinary, straightforward quantum-mechanical calculation and should yield more accurate values about the turn of the century.

LIGAND-FIELD CORRECTIONS

Contributions to the activation energy due to changes in ligand fields are already included, at least formally, in the method of Marcus since the force constants and bond lengths in the activated complex include such changes.

However, the treatment of Hush requires explicit inclusion of ligand-field effects. Hush[15] has suggested the following approach.

The LFSE of the activated complex is approximately equal to the sum of the LFSE for each reactant ion, with the charge and metal-ligand distance characteristic of that ion in the activated complex. Thus, if the activated complex contains two octahedral ions, the LFSE of the activated complex will be given by

$$(\text{LFSE})^{\ddagger} = \frac{[6n_1(e_g) - 4(n_1(t_{2g}) + \lambda)]D_1}{(r_1^{\ddagger})^6} + \frac{[6n_2(e_g) - 4(n_2(t_{2g}) - \lambda)]D_2}{(r_2^{\ddagger})^6}$$

(6–36)

if the transferring electron is of t_{2g} symmetry, and by

$$(\text{LFSE})^{\ddagger} = \frac{[6(n_1(e_g) + \lambda) - 4n_1(t_{2g})]D_1}{(r_1^{\ddagger})^6} + \frac{[6(n_2(e_g) - \lambda) - 4n_2(t_{2g})]D_2}{(r_2^{\ddagger})^6}$$

(6–37)

if the transferring electron is of e_g symmetry. In these equations, $n_i(e_g)$ and $n_i(t_{2g})$ are the number of e_g and t_{2g} electrons of the ith ion; $\lambda = \lambda^{\ddagger}$ is the fraction of an electron transferred from reactant 2 to reactant 1 in the activated complex; $r_1^{\ddagger}$ and $r_2^{\ddagger}$ have the meaning previously assigned to them; and D is, for the indicated reactant, a constant related to the experimentally determined ligand field parameter (Dq) by

$$(\text{Dq})_1 = \frac{D_1}{r_1^6}, \qquad (\text{Dq})_2 = \frac{D_2}{r_2^6}$$

(6–38)

for dipolar ligands or a similar expression involving r_1^5 and r_2^5 for ionic ligands and in which r_1 and r_2 are the equilibrium metal-ligand distance for the indicated ions.

The LFSE of the ground-state reactants must be subtracted from the appropriate expression for $(\text{LFSE})^{\ddagger}$ to obtain the change in LFSE.

A term for spin pairing and spin unpairing should also be added when these quantities are not equal to zero or when they do not cancel.

CROSS REACTIONS

Because the theory for adiabatic electron-transfer reactions cannot be expected to give good absolute values for the rate constants using information now available, it is reasonable to test the theory for correctness of form rather than absolute accuracy. Marcus[10] has developed expressions for the outer-sphere mechanism which, if the theory is correct, should predict the change in rate constant when a single oxidant is used with a series of reductants or which should predict the forward and backward rate constants for a hetero-nuclear electron-transfer reaction when the equilibrium constant of this reaction and the rate constants of the corresponding pair of homonuclear reactions are known. The general procedure is to provide ratios of rate constants in such a way that terms in the activation free energy which are most difficult to estimate reliably cancel. Let us examine the "cross-reaction" test.

Consider the two homonuclear electron-transfer reactions:

$$\text{Ox}_1 + \text{Red}_1 \xrightleftharpoons{k_1} \text{Red}_1 + \text{Ox}_1 \tag{6-39}$$

$$\text{Ox}_2 + \text{Red}_2 \xrightleftharpoons{k_2} \text{Red}_2 + \text{Ox}_2 \tag{6-40}$$

The "cross reaction" is

$$\text{Ox}_1 + \text{Red}_2 \xrightleftharpoons{k_{12}} \text{Red}_1 + \text{Ox}_2; \quad K_{12} \tag{6-41}$$

or the analogous reaction between Ox_2 and Red_1. The relation is

$$k_{12} = (k_1 k_2 K_{12} f)^{1/2} \tag{6-42}$$

in which f is defined so that Eq. 6–42 is obeyed. Thus

$$\log f = \frac{1}{2.303RT} (\Delta F_1^* + \Delta F_2^* + \Delta F_{12}^\circ - 2\Delta F_{12}^*) \tag{6-43}$$

in which the ΔF_i^* are the free-energy changes given by Eq. 6–29 for the rate constants indicated by the subscript and ΔF_{12}° is equal to $-RT \ln K_{12}$ for the medium used. For reaction (6–39) and (6–40), we have

$$\Delta F_1^* = w_{11} + \tfrac{1}{4}\lambda_{11} \tag{6-44}$$

$$\Delta F_2^* = w_{22} + \tfrac{1}{4}\lambda_{22} \tag{6-45}$$

where w has been written for $\Delta F_{c,R}^\ddagger$ in Eq. 6–29, $\lambda = \lambda_0 + \lambda_i$, and $m^2 = \tfrac{1}{4}$. For reaction (6–41) we have

$$\Delta F_{12}^* = w_{12}^R + \left(\frac{1}{4} + \frac{C_2}{2\lambda_{12}} + \frac{C_2^2}{4\lambda_{12}^2}\right)\lambda_{12} \tag{6-46}$$

in which $w_{12}^R = \Delta F_{c,R}^\ddagger$ for reaction (6–41), m_{12} has been expressed as in Eq. 6–31, and C_2, defined by Eq. 6–26, can be written as

$$C_2 = \Delta F_{12}^\circ + w_{12}^P - w_{12}^R$$

Upon substitution of Eqs. 6–44, 6–45, and 6–46 in Eq. 6–43 for $\log f$, it is found that

$$\log f = \frac{1}{2.303RT}\left\{W + \Lambda - \frac{\Delta F_{12}^\circ}{\lambda_{12}}\left(\frac{\Delta F_{12}^\circ}{2} + w_{12}^P - w_{12}^R\right)\right\} \tag{6-47}$$

in which

$$W = w_{11} + w_{22} - w_{12}^R - w_{12}^P - \frac{(w_{12}^P)^2}{2\lambda_{12}} - \frac{(w_{12}^R)^2}{2\lambda_{12}} + \frac{(w_{12}^P)(w_{12}^R)}{\lambda_{12}}$$

$$\Lambda = \frac{\lambda_{11}}{4} + \frac{\lambda_{22}}{4} - \frac{\lambda_{12}}{2}$$

If the assumptions are made that the internuclear metal ion–metal ion distances in the three activated complexes, $X_1^{\ddagger}$, $X_2^{\ddagger}$, and $X_{12}^{\ddagger}$ of Eqs. 6–39, 6–40, and 6–41, respectively, are approximately equal and that the charge on $Red_1(z_1)$ is equal to the charge on $Red_2(z_2)$, then is is found that $W = 0$ and $w_{12}^P = w_{12}^R$ in Eq. 6–47. Furthermore, if it is assumed that

$$\lambda_{12} = \frac{\lambda_{11} + \lambda_{22}}{2} \tag{6–48}$$

it is also found that $\Lambda = 0$. Essentially, Eq. 6–48 means that the contribution of Ox_1 to λ_{12} is the same as it is to λ_{11}, namely, $\lambda_{11}/2$. A similar statement can be made about the contribution of Red_2 to λ_{12}. This is likely to be the case only if the interaction energy in the activated complex is small and if $m \simeq \frac{1}{2}$. For reactions with ΔF_{12}° very large, m will not be approximately equal to $\frac{1}{2}$. When $\Delta F_{12}^\circ \ll 0$, m is very small and ΔF_{12}^* is due only to the coulombic contribution plus a small rearrangement contribution. As a result, the activated complex looks very much like the charged reactants' collision complex. When $\Delta F_{12}^\circ \gg 0$, the activated complex closely resembles the charged products' collision complex. This is in accord with the *principle of similitude*[175] which states that the smaller the free energy of activation the more closely do the molecular details of the activated complex resemble those of the normal reactants. Hence when $\Delta F_{12}^\circ \ll 0$ the contribution of Ox_1 to λ_{12} is very much less than to λ_{11} or, when $\Delta F_{12}^\circ \gg 0$, very much greater. In other words, if the fraction of an electron gained by Ox_1 in the activated complexes of a series of reactions is the same for every reaction in the series, then the rearrangement free energies of the inner and outer coordination shells of Ox_1 are the same in all members of the series. Equation 6–48 would divide the rearrangement free energy of the inner and outer coordination shells in the cross reaction into two independent contributions, one from Ox_1 and one from Red_2 in Eq. 6–41. This approximation would probably be valid for the inner coordination shells of the two reactants because the extent to which these must be adjusted depends on m and not directly on the nature of the reactant partner. Provided a series of reactant partners give the same value of m, the inner-sphere rearrangement energy of Ox_1 would remain approximately constant throughout the series. However, it is seen from Eq. 6–21 that λ_0 for outer-sphere rearrangement consists of three terms, one dependent on Ox_1, one dependent on Red_2, and one term (in $1/r_{\ddagger}$) dependent on the reactants considered together. The latter term will remain constant only if the radii $r_2^{\ddagger}$ of the various reductants, Red_2, remain constant throughout the series.

In particular, Eq. 6–48 means that the deuterium effect for a given reactant, say $Co(ND_3)_5(OD_2)^{+3}$, should be the same throughout a series of reactions, provided the mechanism does not change and the restrictions already noted are met. We have seen in Chapter 3 that this has sometimes been observed for $Cr(H_2O)_6^{+2}$. However, the total deuterium effect on rate constants of a series may vary because the second reactant is continually changing throughout the series, and the deuterium effects contributed by these reactants may vary considerably, depending on the nature of the second reactant.

If the three assumptions stated above are valid, Eq. 6–47 can be written as

$$\log f = -\frac{1}{2.303RT}\frac{(\Delta F_{12}^{\circ})^2}{(\lambda_{11} + \lambda_{22})} \tag{6-49}$$

Using Eqs. 6–44 and 6–45 and assuming that w_{11} and w_{22} are negligible compared to ΔF_1^* and ΔF_2^*, respectively, Eq. 6–49 can be readily transformed into

$$\log f = -\frac{(\Delta F_{12}^{\circ})^2}{(2.303)(4)RT(\Delta F_1^* + \Delta F_2^*)}$$

or, using Eq. 6–5, into

$$\log f = \frac{(\log K_{12})^2}{4 \log (k_1 k_2/Z^2)} \tag{6-50}$$

the usual expression for f.

It is convenient to rewrite Eq. 6–43 as

$$\Delta F_{12}^* + (0.5)(2.303)RT \log f = 0.5(\Delta F_1^* + \Delta F_2^*) + 0.5\,\Delta F_{12}^{\circ} \tag{6-51}$$

If, in a series of reactions of Ox_2 with various reductants Red_1, $\Delta F_2^* \gg \Delta F_1^*$ for all reductants in the series, then

$$\Delta F_1^* + \Delta F_2^* \simeq \Delta F_2^*$$

and a plot of the left-hand side of Eq. 6–51 versus ΔF_{12}° for the series should yield a straight line of slope 0.5 and intercept $0.5\,\Delta F_2^*$. Several such series have been studied by Sutin and co-workers. The series

$$Fe(s\text{-}ph)_3^{+3} + Fe^{+2}(aq) = Fe(s\text{-}ph)_3^{+2} + Fe^{+3}(aq)$$

where s-ph is 1,10-phenanthroline or a substituted 1,10-phenanthroline, was studied by Ford-Smith and Sutin.[87] Other series studied may be written

$$M(z + 1) + Fe(s\text{-}ph)_3^{+2} = M(z) + Fe(s\text{-}ph)_3^{+3}$$

in which $M(z + 1)$ is Ce(IV),[102] Mn(III),[176] and Co(III).[177] When $\log f$ was calculated from Eq. 6–50, the observed slopes in the four series were equal to 0.56, 0.48, 0.49, and 0.51, respectively, in very good agreement with those expected from Eq. 6–51 if the assumptions involved were valid. However, the intercepts were smaller[177] than those predicted. Marcus[10] has suggested that this result may be due to a failure of non-electrostatic terms in the work done in bringing two reactants together to cancel.

TABLE 6–I

Comparison of k_{12} Values Calculated from Eqs. 6-42 and 6-50 with Experimental Values

Reaction	Observed $k_{12}M^{-1}$ sec.$^{-1}$	Calculated[a] $k_{12}M^{-1}$ sec.$^{-1}$
$Ce(IV) + W(CN)_8^{-4}$	$>10^8$	4×10^8
$Ce(IV) + Fe(CN)_6^{-4}$	1.9×10^6	8×10^6
$Ce(IV) + Mo(CN)_6^{-4}$	1.4×10^7	1.3×10^7
$IrCl_6^{-2} + W(CN)_8^{-4}$	6.1×10^7	6.1×10^7
$IrCl_6^{-2} + Fe(CN)_6^{-4}$	3.8×10^5	7×10^5
$IrCl_6^{-2} + Mo(CN)_8^{-4}$	1.9×10^6	9×10^5
$Mo(CN)_8^{-3} + W(CN)_8^{-4}$	5.0×10^6	4.8×10^6
$Mo(CN)_8^{-3} + Fe(CN)_6^{-4}$	3.0×10^4	2.9×10^4
$Fe(CN)_6^{-3} + W(CN)^{-4}$	4.3×10^4	6.3×10^4

[a] The following oxidation potentials and homonuclear electron-transfer rate constants were used in the calculation of the cross-reaction rate constants[177]:

$Ce(III)/Ce(IV)$, 1.44 V, 4.4 M^{-1} sec.$^{-1}$ in 0.4 M H_2SO_4
$W(CN)_8^{-3}/W(CN)_8^{-4}$, 0.54 V, 7 $\times$ 10^4 M^{-1} sec.$^{-1}$
$Fe(CN)_6^{-3}/Fe(CN)_6^{-4}$, 0.68 V, 3 $\times$ 10^2 M^{-1} sec.$^{-1}$ in 0.01 M KOH
$Mo(CN)_8^{-3}/Mo(CN)_8^{-4}$, 0.80 V, 3 $\times$ 10^4 M^{-1} sec.$^{-1}$
$IrCl_6^{-2}/IrCl_6^{-3}$, 0.93 V, 2 $\times$ 10^5 M^{-1} sec.$^{-1}$

In other cases it is more convenient to apply Eqs. 6–42 and 6–50 rather than Eq. 6–51, and this has been done[177] for a number of reactions involving Ce(IV), $IrCl_6^{-2}$, and complex cyanides of iron, tungsten, and molybdenum. The results, listed in Table 6–1, show good agreement between calculated and experimental values, but the agreement may be somewhat fortuitous. For example, the rate constant of the homonuclear electron-transfer reaction between $Fe(CN)_6^{-4}$ and $Fe(CN)_6^{-3}$ was taken to be 3 $\times$ 10^2 M^{-1} sec.$^{-1}$, the value found for 0.01 M KOH,[91] whereas the cross reactions involving $Fe(CN)_6^{-4}$ and $Fe(CN)_6^{-3}$ were carried out in 0.5 M H_2SO_4. However, the rate constant for the cross reaction depends on the square root of the exchange rate constants, k_1 and k_2, so that errors in the latter rate constants produce a much smaller error in k_{12}. Likewise, in cross reactions such as given in the first six entries of Table 6–1, the charges on Red$_1$ and Red$_2$ are unequal, i.e., $z_1 \neq z_2$, and the various electrostatic work terms in W only partially cancel. However, since the rate constants for the homonuclear electron-transfer reactions and the cross reactions are (with the exception of that for Ce(III) + Ce(IV)) rather large, the work terms will be rather small and non-cancellation will be unimportant.

FURTHER REFINEMENT OF THE THEORIES

A complete theory of electron-transfer reactions must include, in addition to a transmission coefficient or collision number, a reasonably accurate treatment of the following:

(a) *Solvent rearrangement.* Levich and Dogonadze used a macroscopic but quantum-mechanical approach via polaron theory. They treated the solvent as a continuum with motional properties described by harmonic oscillator approximations and considered only one oscillator frequency in obtaining their final expressions. They neglected dielectric dispersion of the solvent. Marcus originally used a continuum theory[6] but later provided a statistical-mechanical treatment based on microscopic states.[9] His use of the concept of "equivalent equilibrium distribution" to describe the ligand and solvent distribution in the non-equilibrium state of the activated complex is both illuminating and convenient for calculational purposes.

(b) *Rearrangement of the nuclei of the reactants proper.* Levich and Dogonadze ignore this important process. Hush and Marcus include these rearrangements in different ways and only to the harmonic oscillator approximation though this is probably sufficient.

(c) *Effect of changes in dimensions of reactants on interaction with solvent.* Marcus[11] has recently introduced this factor into his treatment. It was first treated by Sacher and Laidler[16] but does not appear in other treatments. As yet, however, no attempt has been made to introduce correlation between reactant nuclear fluctuations and fluctuations of the solvent.

(d) *Change of vibrational quantum numbers as a means for establishing the energy-balance condition.* Energy may be conserved during the electron-transfer act in different ways, some of which involve energy exchange between vibrational and electronic degrees of freedom with consequent changes in vibrational quantum numbers in forming the immediate products. It is necessary to consider such processes not only when vibronic coupling is significant but also when spontaneous processes involving redistribution of energy in uncoupled electronic and vibrational modes occur.

(e) *Partial dielectric saturation in the solvent.* Laidler[5] considered this problem and Marcus has made some allowance for it. The matter is certainly important in at least some cases and warrants further consideration.

(f) *The collision number for adiabatic outer-sphere reactions.* Up to the present, the collision numbers appearing in the rate expression have been calculated by the Debye method.[173] This procedure does not adequately predict either first collision numbers or cage effects, but its use is unnecessary. There are a number of more detailed treatments of the collision problem in solution as well as measured collision numbers from quenching experiments which can be used for greater accuracy when such accuracy is required (cf. page 120).

(g) *Effects of salt ions in the solvent medium.* Thus far the effect of the Debye atmosphere has been considered only insofar as it influences the coulombic interaction between reactants in the collision complex. This is certainly a major part of the effect of this atmosphere and, in the applications of the theories, this effect has been included by using simple Debye-Hückel corrections. However, the ion atmosphere must also undergo rearrangement about the collision complex as the latter becomes converted to the activated complex and this factor has not often been considered. In other words, there is an additional reorganization of the ion atmosphere accompanying solvent and ligand rearrangement to bring it into equilibrium with Marcus' hypothetical equivalent equilibrium electron distribution of the reactants or with the "real" charge distribution of Hush. The latter two distributions are essentially identical so that the method for including this ion-atmosphere is the same. Formally, it is no more complicated to include this reorganization than to include other reorganization effects, but there is some doubt as to the utility of this correction unless a realistic expression for the ion-atmosphere distribution in terms of reactants' charges is used. Since such expressions are not available, one must have recourse to suitable tabulated data for real systems (a procedure of limited utility) or the alternatives, already mentioned, of carrying out the experiments at very low or very high salt concentrations. However, the problem of treating non-equilibrium polarization of a concentrated ion atmosphere is not a minor one nor can one depend on such high concentrations to make the contribution from such reorganization negligible. This contribution may very well be larger than that due to solvent reorganization. Thus, it seems to be essential that the experiments in electron-transfer kinetics be carried out over a range of low salt concentrations so that extrapolation to infinite dilution is possible.

(h) *Dielectric image forces.* This polarization effect due to charge rearrangement in a dielectric mechanism and the introduction of cavities due to the reactants has been considered only by Marcus. It is probably not of minor importance.

(i) *Summation over all possible activated complexes.* Laidler[5] first made this summation. Levich and Dogonadze and Marcus now make such summations in approximate ways; the method of Marcus is to be preferred. The method is given in Chapter 5 (see Ref. 11).

(j) *Kinetic energy in interreactant coordinate* R. Regardless of the nature of the mixing perturbation and the nature of the reaction coordinate motion, the kinetic energy in R should be taken into account in determining the energy-balance conditions. It is ignored by Levich and Dogonadze and Hush but has been considered by Marcus in his treatment of the reaction coordinate and in the construction of the total system Hamiltonian operator for the activated complex.[178]

(k) *Provision for non-spherically symmetrical reactants.* This matter is undoubtedly of considerable importance and bears on both the problem of counting activated complexes and the calculation of total rearrangement energies. In inner-sphere activated complexes, it should not be difficult to include. For outer-sphere activated complexes, considerable detailed information is required and the treatment is difficult under any circumstances. This factor has already been discussed in minor detail earlier in this chapter.

(l) *Tunnelling of solvent and reactant nuclei.* At ordinary temperatures, tunnelling of solvent nuclei is unimportant. As discussed in Chapter 8, tunnelling processes of reactant nuclei are unlikely to result in a significant decrease in activation energy unless H'_{RP} is large. Levich and Dogonadze have provided a low-temperature rate-constant expression which suggests that tunnelling effects may become important. Marcus[19] has discussed this treatment, but there are as yet no experimental observations to test its validity.

(m) *Detailed treatment of the reaction coordinate.* It appears certain that the reaction coordinate in most electron-transfer reactions is quite complicated. Even if this were not the case, it is necessary to know under what conditions the motion in this coordinate may be considered independent of other motions and particularly under what circumstances the Schrödinger equation can be factored to separate this coordinate from all others. This is a fundamental theoretical problem affecting both enthalpy and entropy calculations. Marcus has recently initiated a series of papers dealing with these topics.[178] The problem is of importance in relation to the approximations of absolute rate theory which implicitly or explicity underlie all the theoretical expressions for the rate constants of electron-transfer reactions. However, it is doubtful that a more detailed analysis of the problem will force significant changes.

(n) *Treatment of the special problems presented by inner-sphere activated complexes.* All theoretical treatments of rate constants thus far presented are explicitly developed for reactions proceeding through outer-sphere activated complexes. The special problems associated with the treatment of inner-sphere reactions have already been mentioned in this chapter. If the rate is limited by electron transfer in the activated complex, the mixing perturbations are probably of the same type as for outer-sphere non-adiabatic reactions, and the problem is slightly simplified by the fixed geometry which limits the range of the internuclear coordinate and reduces the number of acceptable activated complexes. However, specific information about the geometry of the activated complex is required for any given case, the reactant ion-reactant ion interaction becomes considerably more complicated since simple electrostatic expressions may be inadequate, and the ligand motions become more complicated.

The several factors of major importance in this list have been emphasized in this chapter. Marcus,[11] Sutin,[18] and others have estimated the importance of some of the remaining factors and have provided approximate methods for inclusion of the more important factors in this group in the rate constant expression, though no one has considered inner-sphere reactions as a special case. The remaining factors have not yet been considered in sufficient detail to allow an estimation of their importance. It is doubtful, however, that there are any major omissions in the latest version of the theory of Marcus.[11] We have noted that there is relatively little difference between the results of Marcus, Hush, and Levich and Dogonadze. The virtue of the treatment of Marcus lies both in its simplicity and in the fact that it is at present considerably more comprehensive than the others in dealing with the various factors. We have also been forced to conclude that, in those small details in which his rate expression deviates from those obtained by Hush and by Levich and Dogonadze and in certain physical interpretations of important parameters, Marcus is usually more correct. Hence from the practical point of view, there is little reason to emphasize the other treatments. This last statement is not meant to imply that the treatment of Marcus is complete or that all his approximations are satisfactory. In particular, the potential-energy functions for ligand distortion are not yet satisfactorily estimated. Nevertheless, the agreement of prediction and theory is sufficiently good that there is unlikely to be any serious shortcoming in the theory when applied to adiabatic outer-sphere reactions if the model treated contains all the important chemical details of the reaction of interest. For very fast reactions (which have small free energies of activation), it appears that some rate constants can now be predicted within a factor of perhaps ten, i.e., free energies of activation can be predicted within an error of approximately 1.3 kcal./mole at ordinary temperatures. For slow reactions, the agreement between calculated and experimental free energies of activation is reasonably satisfactory and will improve with applications of existing, more refined theoretical developments; if a large discrepancy is found in any particular reaction, one may suspect that the assumed mechanism is in error or that very inaccurate values have been chosen for various parameters appearing in the rate expression. On the other hand, agreement between the calculated and experimental free energies of activation neither implies a correct theory nor that the mechanism is an outer-sphere activated-complex mechanism. Such agreement can, at most, only imply that the outer-sphere mechanism may be operative.

Comparison of calculated and experimental free energies of activation is only a part of the ultimate test of the validity of a theory applied to any particular reaction; frequently it has been the only test applied. A more rigorous test of a theory is comparison of calculated and experimental enthalpies and entropies of activation. Enthalpies and entropies of activation usually cannot be predicted with confidence for reactions between small ions because the temperature dependence of the dielectric constant in the neighborhood of these ions is not understood. When the ionic reactants and products are large, it should be possible to use the temperature coefficient of the bulk dielectric constant, or some reasonably simple modification based on partial dielectric saturation of solvent, to calculate the electrostatic contribution (mainly in $\Delta S_c^{\ddagger}$) to the entropy of activation with much greater confidence. The enthalpy of activation can then be calculated in the usual way by combining ΔF^{*} and ΔS^{*} terms; thus the major contributions of ΔH^{*} will be given by

$$\Delta H^{*} = (\Delta F_c^{\ddagger} + T\Delta S_c^{\ddagger}) + \Delta F_i^{\ddagger} + (\Delta F_0^{\ddagger} + T\Delta S_0^{\ddagger})$$

Expressions for all terms appearing in this equation have been given previously in this chapter.

At present, it is not known how well this more rigorous test is satisfied by the theory. In the case of the $Fe^{+2}(aq) + Fe(ph)_3^{+3}$ reaction discussed in Chapter 3, it is found that the calculated and experimental values of $\Delta F^{\ddagger}$ are in reasonable agreement but that the calculated $\Delta H^{\ddagger}$ and $\Delta S^{\ddagger}$ values are in complete disagreement with the experimental values. Substituting the value of 10.5 kcal./mole calculated for ΔF^{*} in Eq. 6–2, it is readily found that $\Delta F^{\ddagger}$ is 11.3 kcal./mole. However, experimentally it was found that $\Delta F^{\ddagger}$ is nearly equal to $T\Delta S^{\ddagger}$, $\Delta H^{\ddagger}$ being approximately 0.2 kcal./mole, whereas the calculated $\Delta F^{\ddagger}$ value is almost equal to $\Delta H^{\ddagger}$. This difference between calculated and experimental values suggests that the mechanism of this reaction probably does not involve an outer-sphere activated-complex. It is not known with any certainty, however, just how well the theory can predict entropies of activation. Studies should be made of the effect of ionic strength on the values of $\Delta S^{\ddagger}$ for known outer-sphere activated-complex mechanisms. The contribution of $\Delta S_c^{\ddagger}$, given by Eq. 6–7, varies widely with changing ionic strength for reactions between highly charged reactants. For example, when $r_{+} = 7$ Å and $z_1 z_2 = +6$, $\Delta S_c^{\ddagger}$ changes from about -2 cal./deg. mole at $\mu = 1$ to -16 cal./deg. mole at $\mu = 0$. When $z_1 z_2$ is negative, $\Delta S^{\ddagger}$ will increase, instead of decrease, with decreasing ionic strength. Thus it may be possible to test the applicability of the theory to a given reaction on the basis of the ionic strength dependence of $\Delta S^{\ddagger}$ to provide a more rigorous test than the comparison of calculated and experimental $\Delta F^{\ddagger}$ values. Much obviously remains to be done to test, correct, and expand the theories. By a suitable choice of reactions, important questions concerning adiabaticity and non-adiabaticity, reliability of the enthalpy and entropy of activation, and the like may be answered and the answers incorporated into a comprehensive description of at least some electron-transfer reactions. Even at present, the theory for understanding of adiabatic homogeneous electron-transfer reactions is undoubtedly superior to the theory and understanding of other types of chemical reactions.

It is important to remember that the theories are quite general in their applicability to electron-transfer reactions. They are quite as suitable for electron-transfer reactions between organic species, between metal complexes and organic molecules, or between non-metallic inorganic molecules as they are for reactions between complex ions. The last category of reactions has been emphasized, because it is the simplest type for exposition and because such reactions have been studied systematically in ways most useful for comparisons with theoretical predictions. Much less application of electron-transfer theories has been made to the other types of reactions but such applications will become an important area of research in organic chemistry, in particular, within a short time.

7

Heterogeneous Electron-Transfer Reactions

The transfer of electrons between a metal or semiconductor and a dissolved or surface-bound reactant is not different in kind from homogeneous solution processes emphasized in this monograph. Generally speaking, such heterogeneous reactions are more complicated and, invariably, they require more specific information in order to set up a model for theoretical calculations of rate constants. It is also true that there exists a far smaller body of experimental results suitable for testing such calculations and, often, when a set of experimental results appears suitable for such purposes, actual knowledge of the experimental state of the system is so incomplete that a poor comparison may be due to unknown experimental factors just as frequently as to theoretical inadequacies. For such reasons, we shall devote little space to heterogeneous reactions and essentially none to experimental results. Discussion of experimental results can be found in references.[179,180,181] It is useful, however, to consider the special problems requiring treatment in theoretical investigations and to show how closely related are the rate expressions obtained for heterogeneous and homogeneous reactions, using the approximate theories already discussed.

The complications in theory associated with metal electrodes are somewhat less than those associated with semiconducting electrodes, as we shall see. Marcus[11,182] and Hush[183] have both modified their theories for heterogeneous reactions, with emphasis on adiabatic reactions at metal surfaces. Dewald[184] has considered semiconducting electrodes from essentially the point of view of Marcus, and his discussion of the relevant problems of electron distribution in semiconductors should be referred to since we shall not delve deeply into this subject. Gerischer[185] has given an extensive and illuminating discussion of the problem from the point of view of Gurnee. Gerischer gives an even more detailed exposition of the special problems of semiconductors than Dewald and should be read in conjunction with the Dewald papers. By current standards both treatments for rate-constant calculation are incomplete in important ways.   Similar comments apply to the theory of Randles,[186] who first introduced the Franck-Condon restriction for electrode reactions.

Dogonadze and Chizmadzhev[187] have extended the original theory of Levich and Dogonadze[13] to both metals and semiconductors retaining the non-adiabatic basis of the original theory.   There is some consideration of the special problems of semiconductors but the treatment is quite incomplete. The most recent version of Marcus' treatment has been developed more extensively than the others but still remains limited in its inclusion of special complications appearing in heterogeneous reactions.   Before presenting the mathematical results of some of these treatments, the complications will be detailed.

A list of special problems appearing in heterogeneous electron-transfer reactions is as follows:

(a) *Image forces.*   With semi-conducting electrodes in states containing small numbers of conduction electrons, the electrostatic interaction between solid and solution reactant is a local effect similar to the electrostatic interaction between charged reactants in homogeneous reactions.   There is an image force interaction between solution reactant or product and the solid electrode, but it is a dielectric image interaction of the type previously discussed for reactant-solvent interaction in homogeneous reactions and thus is a minor contribution to the activation free energy.   As the population of conduction electrons increases the image force becomes more important, but its importance is determined by the degree to which the conduction electrons or holes resemble the conduction electrons of metals; thus the effect is time-dependent and may be complicated.   In fact, it can be imagined that the changes in the semiconductor required to set up the image is the reaction coordinate in some cases.   That is, in such cases we can think of the re-arrangement of charge carriers in the semiconductor as being similar to the re-arrangement of solvent and ligands.   Alternatively, rearrangements in the semiconductor may have to be treated as separate elementary steps of the over-all reactions if they are very slow, e.g., diffusion processes.   For metals, on the other hand, the problem is much simpler since the velocities of charge carriers are fast with respect to the lifetime of the activated complex.   Then the principal electrostatic interaction in the activated complex is between solvated reactant or product and its image in the metal.   The image force is determined by the dissolved molecule so that the total polarization rearrangement in the activated complex is determined by a single molecule.

(b) *Height and distribution of energy levels in the solid.*   The electron transfer is between the solid surface layer and the dissolved reactant. Beneath this surface layer, for metals, lies a constant potential solid in which no electron localization is possible.   If there is no electrical contact between solution and metal through a reference electrode, i.e., a complete electro-chemical cell, the electron-transfer reaction will proceed to alter this potential by electron addition or removal until equilibrium is obtained. As usual, if the reaction can be followed by isotope exchange in the dissolved reactant, conditions for analysis are simplest, since among other things the over-all free-energy change is zero. However, most heterogeneous reactions must be or have been studied under conditions of constant overvoltage, i.e., with the solid maintained at a potential which is not the true equilibrium value. This potential contributes directly to the over-all standard free-energy change and usually plays a kinetic role as well. A particularly complex example of the kinetic role occurs whenever the reaction at the surface involves surface contamination or, as in the well-known overpotential for oxygen reactions on metals, high-energy intermediates during electron-transfer steps. In any event, under conditions of constant bulk metal potential there is no change in the metal during the reaction, i.e., no conversion from a reactant metal state to a product metal state. The free-energy difference for the over-all reaction is constant. Hence if the electrode reaction is adiabatic, all metal electrodes will be equivalent when adjusted to a potential such that the over-all free energy is the same to a first approximation. The extent to which this is not true will depend on variations in surface roughness, double layers, and contamination, to be discussed below.

In a finite piece of metal the energy levels are very closely spaced. We must immediately recognize that this means a wide distribution of activated complexes, each with a different microscopic over-all free-energy change determined by a specific small interval of energies in the solid. The number of each of these must be weighted by a distribution function for electrons in the solid and by a transmission coefficient. From the theoretical point of view this is unfortunate, since we seldom have much information about the distribution function and the problem of the transmission coefficient for energy levels in a closely spaced array has not been considered in much depth and is probably quite different from the problems of the transmission coefficient for homogeneous reactions. It is thus necessary to work theoretically with averages dependent on simpler transmission-coefficient expressions and known parameters such as the work function of the metal.

The surface energy levels of semiconductors are even more complicated. That part of the over-all potential change which occurs in the solid is distributed over a wide layer below the surface, called the diffuse layer, and jointly with the solution over the Helmholtz layer. The potential drop across the diffuse layer depends on the solution composition and on the steady-state current, since the steady-state current determines the proportions of charge carriers in the diffuse layer. Furthermore, the dependence on the effective potential in the solid is complicated, since the number of charge carriers is determined by this potential. At some lower limit there are few holes and electrons in conduction bands and the solid behaves as a dielectric, though not a simple dielectric, because of the effect of temperature fluctuations on the populations of conductors.   At higher potentials there is a distribution of energy levels in the conduction band similar to the distribution in metals.   The effect of this complication and the effect of solution composition on semiconductor surface potentials needs to be discussed in some depth to be useful.   The discussions of Gerischer are particularly helpful.

(c) *Surface potential gradients.*   There is usually a large potential drop between the surface of a metal and a solution.   In the case of semiconductors a drop is still present, and for the same reasons, but the largest part of the potential drop occurs across the diffuse layer.   The total surface layers extending out from the metal surface to some point at which the solution has uniform bulk properties is usually divided into the narrow Helmholtz layer at the surface and the deeper Gouy layer though this is not a strictly correct definition of the Gouy layer.   The Gouy layer has a Debye length determined by the salt concentration in solution.   At high ionic strength this layer is compressed down into the Helmholtz layer and has an effective thickness of about one ionic diameter.   The effective thickness of the Gouy layer is also determined by the rate of stirring, since it can be sheared by flowing solution.

The two layers are interrelated and depend on a number of factors which are unique to a particular system as well as on the overpotential of the metal. For example, there is usually specific adsorption of solvent ions of one kind of charge.   This is normal behavior and often can be characterized by specific studies of the double layer.   However, the activated complex for electron transfer is formed in these layers.   Its properties will depend on the layers, and it will also perturb the layers.   There is thus introduced a polarization adjustment problem of considerable complexity.   If this problem is a serious one, its analysis must be highly approximate and will lead to large numerical uncertainties in the prediction of rate constants.   Even uncharged species may interfere with these layers.

(d) *Surface adsorption of reactants and products.*   The line between chemisorption and physical adsorption is not always sharp.   Strong chemisorption corresponds to a conventional chemical reaction of strong interaction and primary bond rearrangement in the activated complex and falls outside this discussion.   Somewhat weaker chemisorption probably can be classed with homogeneous electron-transfer reactions passing through inner-sphere activated complexes and thus presumably can be treated on the general framework of electron-transfer theory though, of course, with considerable added difficulty.   Physical adsorption can thus be defined for present purposes as being the heterogeneous analogue of a homogeneous outer-sphere reaction.   Only this type of reaction need be considered at present.

If physical adsorption or desorption is sufficiently slow, the over-all reaction rate will be limited by such processes.   On the other hand, if the reaction is limited by the collision process at the surface or by the electron-transfer act, the thermodynamic contributions from physical adsorption appear as usual in the activation free energy, etc., without specific reference to adsorption as a separate phenomenon. We refer the reaction as usual to standard states in the bulk solution. Physical adsorption of ions or reactants with permanent dipole moments is strongly dependent on electrostatic interactions which are especially favorable for metallic solids. On the other hand, van der Waals forces and hydrophobic-bonds forces (polar–non-polar surface activity effects) depend on the character of the solvent as well as the polarizability of the solid. In particular, we expect non-polar parts of the reactant and product to find favorable location at the surface of a semiconductor if the solvent is water. In the case of a metal, this hydrophobic situation is less clear because of the image force interaction between water molecules and the metal. Similarly, reactants of high polarizability will be absorbed on the metal surface from the bulk solvent more readily in solvents of low polarizability than from solvents of high polarizability. Undoubtedly, these interactions often make important contributions to the free energy of activation. A more important factor, in general, will be the favorable or unfavorable manner in which the reactant and product fit into the double layer, as has already been discussed.

(e) *Surface adsorption of contaminants.* A clean solid surface usually attracts a reasonably complete monolayer of one of the solution components. Many metals form oxide or hydrous oxide layers. The surface-bound material is a contaminant if it is not the reactant, and there arises the question as to whether the electron-transfer reaction occurs at uncontaminated surface sites, by displacement of contaminant, or between the contaminant and the dissolved reactant. All three alternatives require special consideration in choosing a model and hence in activation energy calculations. Each case is different and one must not only know *a priori* which is correct but also be able to make quantitative estimates of physical or chemical adsorption free energies, enthalpies, and entropies. To further complicate the picture there is the problem of surface roughness, which has both geometric and chemical-potential aspects, and the general problem of surface defects. Unsaturated valences on the solid surface due to roughness, inadequate annealing, etc., can strongly influence adsorption phenomena, surface electron distribution and, presumably, transmission coefficients.

(f) *The collision number.* For adiabatic reactions, collision numbers are required. These will depend on the number of solid sites for activated complex formation, the physical description of such sites if the surface is rough, and the solution flow near the surface. If there is a pre-equilibrium surface adsorption step or steps, the model must be adjusted as required to take account of such steps. Reactions in which surface adsorption or desorption is rate-limiting require a different treatment and are trivial from the electron-transfer point of view but obviously not trivial from the point of view of mechanisms.

Usually it will be difficult to calculate collision numbers but there is considerable relevant information on such collision processes, and there are a variety of methods which may be used to provide an empirical estimate of this number, though preferential reaction at a limited number of surface sites and adsorption processes can limit the validity of such estimates.

At present we have no very good estimate of the importance and occurrence of non-adiabatic, heterogeneous, electron-transfer reactions. It is probable that they are even less important than non-adiabatic homogeneous electron-transfer reactions. Most theoretical treatments have assumed adiabaticity or have provided some sort of semiquantitative argument against non-adiabaticity. Much complication and need for special information about surface sites, etc., is eliminated in adiabatic reactions, but the calculation of the free energy of activation still remains difficult.

The theories of heterogeneous electron-transfer reactions presented thus far ignore most of the specific complications listed above which require detailed information about the surface and solid. We shall confine ourselves to brief summaries of the treatments given by Levich *et al.* and by Marcus.

THE THEORY OF MARCUS

This is an adiabatic theory with transmission coefficient of 0.5. It has been developed for metal electrodes. It could be modified for a non-adiabatic reaction by elimination of the collision number and the use of a transmission coefficient of the L.Z.S. form and an integration over some approximate distribution of activated complexes according to energy level and electron distributions in the metal. The latter modifications are not simple nor general, and it is not clear that the modifications possible on the basis of current information would lead to accurate expressions for the rate constant.

The most recent version of the theory treats solvent polarization by the statistical-mechanical method of the equivalent equilibrium distribution and includes ligand reorganization in a quadratic approximation. Motion in the reaction coordinate was treated classically. These aspects of the treatment are identical in homogeneous and heterogeneous reactions. For metals the energy levels making major contributions to the rate are considered to lie within kT of the Fermi energy. The latter is the energy of the highest filled energy level at absolute zero. Then, in general, the sum over activated complexes is replaced by a single activated complex with properties corresponding to the Fermi level.*

* Marcus has included a factor, ρ, multiplying the right hand side of Eq. 7-1 to include the fact that not all the range of interreactant (surface to reactant) R values consistent with the equivalent equilibrium distribution may have corresponding activated complexes making important contributions to the reaction; this is a counting

Thus the rate constant is given by

$$k = Z\kappa e^{-\Delta F^*/kT} \tag{7-1}$$

in which Z is the collision frequency for an uncharged reactant with a unit area of electrode surface, ΔF^* is given by Eq. 7-2, and κ is a transmission coefficient.

$$\Delta F^* = \frac{w_R^* + w_P^*}{2} + \frac{(\lambda_i + \lambda_0)}{4} + ne\frac{(E - E_0')}{2} + \frac{(neE - neE_0' + w_P - w_R)^2}{4(\lambda_i + \lambda_0)} \tag{7-2}$$

Z can be approximated as a one-dimensional velocity, $\sqrt{kT/2\pi\mu}$, if the surface is smooth and uniform and the reactant unrestricted in its motion parallel to the surface. The transmission coefficient has been taken as 0.5, but more generally it would have the appropriate value for the Fermi level weighted for velocity in the reaction coordinate and for the distribution of holes or electrons in the metal, depending on whether the metal gains or loses electrons in the reaction. The transmission coefficient is also dependent on over-potential.

In Eq. ·7–2, E is the total electrical potential difference between solution and electrode under the conditions of the experiment, and E_0' is the corresponding potential at chemical equilibrium. The difference is the over-voltage. The work required to bring the reactant to collision-complex distance (see Chapter 5) is w_R and includes effects of electrical forces, van der Waals forces, "hydrophobic forces," as well as work done against chemical potential and electric forces of the Helmholtz layer and the electrical forces of the Gouy layer. w_R might be thought of as the adsorption energy of the reactant. The work term w_P is the corresponding quantity for the product. The "equivalent-equilibrium" parameter λ_i is equal to one-half of the value for the corresponding homogeneous homonuclear electron-transfer reaction. The one-half factor appears because the metal effectively displaces one of the reactants in the homogeneous phase reaction. The equivalent equilibrium parameter λ_0 takes into account the solvent polarization by the image force and is equal to one-half its value in the corresponding homogeneous homonuclear electron-transfer reaction at an internuclear distance equal to that between the reactant and its image in the actual heterogeneous reaction. It also includes ligand reorganization and most refinements in the theory will

---

correction to the statistical-mechanical procedure. However, the form of this correction factor is such that formally, at least, it may also be considered a collision-number correction or "steric factor." Thus cage effects of solid or Gouy layer, surface roughness as it determines number of sites for activated-complex formation, surface contamination insofar as it reduces available surface area, and the counting of effective numbers of sites on a non-uniform (in chemical potential or transmission coefficient) surface can be included in this steric factor. Unfortunately, this formal simplification does not simplify the estimation of these effects.

appear in the calculation of these quantities, i.e., in increasing refinement of the total "equivalent-equilibrium" condition. Modification due to non-spherical reactants appear in these parameters though it could be accommodated by a different summing procedure in the counting of activated complexes.

Modifications of the theory for use with semiconducting electrodes can be made on the basis of the present formal framework. That such modifications likely to provide reliable expressions are difficult to treat precisely must be apparent from what has already been written. The first steps taken in this ambitious direction by Dewald are presented in his papers. It is not yet possible to make confident comparisons of experimental and theoretical predictions for semiconducting electrodes.

THE THEORY OF DOGONADZE AND CHIZMADZHEV

This is a non-adiabatic theory following the homogeneous reaction theory of Levich and Dogonadze step by step. The solvent is treated macroscopically and as a continuous medium outside the inner coordination shell. Ligand rearrangement has been ignored. Both semiconductor and metal electrode cases have been examined. In both cases the problem is treated in terms of a "one-electron" wave function approximation for the migrating electron in the solid, reactant and product. The energy levels of the solid are treated individually. Specific attention is given to the double layer which is assumed to be treated by a self-consistent field approximation. The perturbation of the double-layer field by the reactant or product ion is shown to be a small effect in at least some situations.

For small transmission coefficients, using the notation of Marcus, the reaction probability at R is given by

$$P_{RP} = \sqrt{\frac{\pi}{kT\lambda_0}} \frac{|H'_{RP}|^2}{\hbar} \exp\left[-\frac{(\Delta E + \lambda_0)^2}{4\lambda_0 kT} \right] \qquad (7\text{--}3)$$

in which H'_{RP} is the usual interaction matrix element (cf. Chapter 5) and ΔE is the over-all standard electronic-energy change in the reaction but at collision-complex dimensions rather than at infinite separation of reactant (or product) and electrode surface. The parameter λ_0 is assumed to be identical with the λ_0 parameter in the homogeneous reaction though this must be considered a rather poor approximation.

If the density of energy levels in the electrode is σ and if the electron distribution probability function is $n(E)$, the sum over probabilities is given by

$$k_{RP} = \int_{R=0}^{\infty} \frac{C_2(R)}{C_2(\infty)} dR \int P_{RP}(R)[1 - n(E)]\sigma \, dE \qquad (7\text{--}4)$$

in which $C_2(R)$ is the equilibrium concentration of reactant at R and $C_2(\infty)$

that at $R = \infty$. Since P_{RP} increases rapidly near $E = E_F$ and $n(E)$ decreases rapidly in the same region E may be set equal to E_F in evaluating the integral. Then the forward and reverse rate constants are approximated by

$$k_{RP} = \sqrt{\frac{\pi kT}{\lambda_0}}\, \sigma_F \frac{|H'_{RP}|^2}{\hbar} e^{-w_R/kT}\, \delta R \, \exp\left\{-\frac{[w_P - w_R + (E - E'_0) + \lambda_0]^2}{4\lambda_0 kT}\right\}$$

(7-5)

$$k_{PR} = \sqrt{\frac{\pi kT}{\lambda_0}}\, \sigma_F \frac{|H'_{RP}|^2}{\hbar} e^{-w_P/kT}\, \delta R \, \exp\left\{-\frac{[w_R - w_P - (E - E'_0) + \lambda_0]^2}{4\lambda_0 kT}\right\}$$

(7-6)

The quantity δR is defined on page 112; E, E'_0, w_R, and w_P have the same significance as in the heterogeneous reaction expressions of Marcus.

The well-known Frumkin equation for the exchange current as a function of concentration and overpotential has been derived as a limiting case under rather severe assumptions by both Marcus and Dogonadze and Chizmadzhev.

As written, Eqs. 7-5 and 7-6 apply to metal electrodes. For semi-conductors valence and conduction electrons are treated separately and some of the several problems associated with the change in electrical potential through the Helmholtz and diffuse layers were considered. This involved replacing the over-all free-energy change, when the reactants are at R, $[w_P - w_R + (E - E'_0)]$, by an expression appropriate to the properties of the semi-conductor.

The theory of Dogonadze and Chizmadzhev can be modified for adiabatic electrode reactions in the same way the homogeneous reaction theory was modified by Dogonadze. The criticisms of the theory are the same as we have previously given for the homogeneous non-adiabatic theory.

In general form, there appears to be little improvement to be asked in the formal theories of either Marcus or Dogonadze and Chizmadzhev. The application of these theories is quite another matter since it is obviously very difficult to estimate a number of the parameters appearing in the reaction-rate expressions. The problems of providing absolute rate constants are then problems of estimating the values of these parameters, an undertaking apparently more formidable than the analogous undertaking for homogeneous reactions. For the present we must be less ambitious and look for the consistencies among series of reactions, predicted from the theoretical expressions to provide criteria for the adequacy of the theory. By suitable choice of comparisons factors of unknown numerical value can be caused to cancel. Eventually this type of correlation may actually provide means for empirical evaluation of these values. Much more experimental work is required but even with the fragments of data now available there are some encouraging indications that the theoretical expressions are correct in form for the factors they include even though they are still incomplete. The more important theoretical predictions and the limited experimental information relevant at this time to the testing of these predictions are discussed in some detail by Marcus[19] but it is worth summarizing his discussion as the conclusion to this chapter in order to demonstrate the fundamental unity of all electron-transfer mechanisms.

(a) The quantity α, called the electrochemical transfer coefficient, in

$$k_{\text{cathodic current}} = k^\circ_{\text{cathodic}}\, e^{-(\alpha n \mathscr{F}/RT)\varepsilon}$$
$$k_{\text{anodic current}} = k^\circ_{\text{anodic}}\, e^{(1-\alpha)n \mathscr{F}\,\varepsilon/RT} \tag{7-7}$$

is 0.5 for reactions at metal electrodes so long as the free-energy contribution due to the overpotential is small with respect to the activation free energy for the exchange current, i.e., $|m\mathscr{F}(E - E'_0)| < |\Delta F^*|$. This generalization must be qualified, however, to the extent that it applies only when the work terms in forming the collision complex from reactants or products are either small or, if large, have been removed from the total free energy of activation by an independent procedure.

The exchange current is the current flowing in either direction when the electrochemical cell is at equilibrium. In Eq. 7-7 ε is the electrode potential usually referred to the standard hydrogen electrode, $k^\circ_{\text{cathodic}}$ and k°_{anodic} are rate constants at $\varepsilon = 0$ and $\mathscr{F}$ has its usual meaning.

(b) Furthermore, when a series of reactants is oxidized (or reduced) at a single kind of metal electrode under such conditions that there is a constant potential difference between metal and solution, the ratios of the electrochemical rate constants to the corresponding chemical rate constants obtained from the same series using a common reductant (or oxidant) in homogeneous solution are all the same. This is true if the electrochemical transfer coefficient is 0.5 for the electrode reactions and if the slope of the free energy of activation versus standard free-energy change for the homogeneous reactions, the "chemical transfer coefficient," is 0.5. Again the work terms must be small or must be removed before making the comparison.

(c) When the collision-complex work terms are small, the rate constant, k_{ex}, for a homonuclear electron-transfer reaction in solution is related to the rate constant, k_{el}, of the exchange current in the electrochemical reaction of the same two oxidation states at a metal electrode by

$$\left(\frac{k_{\text{ex}}}{Z_{\text{soln}}}\right)^{1/2} \simeq \frac{k_{\text{el}}}{Z_{\text{el}}} \tag{7-8}$$

The collision numbers Z_{soln} and Z_{el} refer to the chemical and electrochemical reactions, respectively. These approximate relations and others are derivable in the manner shown on page 133 (see Ref. 10 for details).

Experimental results, though limited, are in good agreement with the above predictions. The electrochemical transfer coefficient for a number of reactions is 0.5 as predicted in (a). Prediction (c) is also correct for a number of examples[10] and some of the more glaring exceptions are explicable on the basis of the more detailed knowledge of the chemical reactions now available. The predictions are strictly applicable only to outer-sphere reactions and indeed provide a fair test to distinguish between outer-sphere and inner-sphere mechanisms. When comparisons are made using only outer-sphere cases, prediction (b) comes out well. For example, when the rate data for the series of chemical reactions of $Co^{III}(NH_3)_5X$, with various X, and the reductants Cr^{+2}, Eu^{+2}, V^{+2}, $Cr(bipy)_3^{+2}$ (see Ref. 95), or $Ru(NH_3)_6^{+2}$ (see Ref. 94) are compared with the corresponding electrochemical rates for some members of the series measured at a dropping mercury electrode,[188] the chemical and electrochemical rate constants for a given $Co^{III}(NH_3)_5X$ form a set of ratios which tend to remain the same for other X except in cases for which inner-sphere mechanisms are thought to be involved. The agreement is not perfect especially for homogeneous reduction by Eu^{+2} but it is good enough to establish the basic similarity of chemical and electrochemical reactions. Thus for at least outer-sphere reactions, the reaction model and the consequent theoretical formulations appear to be correct though still lacking in some details. Future theoretical work will thus be built on the present base and concern itself with the inclusion of details, treatment of inner-sphere mechanisms and the problem of non-adiabaticity.

8

Nuclear Tunnelling

Except in a few cases, the energy-balance condition in the activated complex cannot be reached without displacements of at least some of the coordinates of ligands or solvent into ranges of the coordinates lying outside the normal oscillatory ranges characterizing ligand vibration and solvent rotation and vibration in the ground state of the collision complex. Whether or not these distortions ultimately lead to changes in quantum numbers for these modes in the products' activated complex is essentially irrelevant since, insofar as we have been able to estimate, the activated complex has too short a lifetime for the concept of stationary vibrational states in these modes to have any meaning. In any event, most of the activation energy is associated with these nuclear distortions to which we also look for the mixing perturbations. It has been suggested by Sutin[18] that the activation energy can be reduced if it is possible for some of the nuclei thus affected to pass from the equilibrium geometry in the collision complex for reactants to their equilibrium geometry in the products' collision complex by an efficient tunnelling mechanism which will take the nuclei through the restrictive potential barriers rather than over them. This could be an important detail of electron-transfer reactions, and the question we must try to answer is just how important such processes actually are.

In the beginning, we note from Fig. 8–1a that for non-adiabatic reactions there can be no nuclear tunnelling since there is extremely little mixing of zero-order states and thus no "two-sided" potential barrier. It may be said that a two-sided barrier must exist since the states mix in electron transfer, but the range of coordinate positions on either side of the activated complex region in which this barrier exists is trivially small so that nearly complete nuclear distortion has to be accomplished before a "two-sided" barrier exists and tunnelling can occur. Imagine a point, representative of the nuclear configuration of the reactants in a non-adiabatic reaction, moving along the potential-energy surface of the reactants in phase space. Everywhere on this surface, the mixing of the zero-order reactant and product states is so small that there is essentially only one potential-energy surface for the system of nuclei, namely, the zero-order surface of the reactants. As a result, the potential-energy barrier is virtually of infinite width (except in the activated-complex region) and no tunnelling of the nuclei can occur.

On the other hand, in adiabatic reactions the zero-order states become mixed as the reacting system approaches the activated-complex region. The system in this region is moving on the lower first-order state produced by mixing. A two-sided potential barrier for nuclear motions now exists and

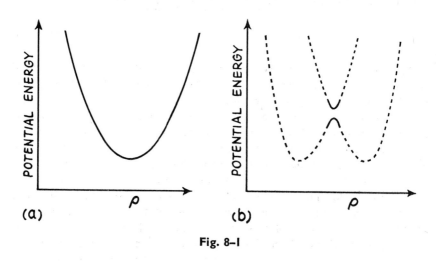

Fig. 8–1

the nuclei of the ligands or the solvent can "move" through this barrier. The transmission coefficient for this tunnelling process will depend on the height of the barrier, its width, *and* the extent of mixing of the zero-order surfaces. If the mixing interaction is weak, the transmission coefficient for nuclear tunnelling can be appreciable only very near the top of the barrier. If the mixing interaction is strong, i.e., if H'_{RP} is quite large, the transmission coefficient for nuclear tunnelling may become significant both because the height of the potential barrier is lowered and because the large interaction between zero-order states allows a change of electronic configuration to accompany the nuclear tunnelling. Thus the transmission coefficient depends on H'_{RP} in two ways, i.e., through its effect on the barrier height and through its control on the extent of mixing. The tunnelling probability rises very rapidly with H'_{RP} and we may expect nuclear tunnelling to be much more important in "strongly adiabatic" (large H'_{RP}) reactions than in "weakly adiabatic" (smaller H'_{RP}) reactions and unimportant in non-adiabatic (very small H'_{RP}) reactions. In strongly adiabatic reactions, the reaction path may involve passage through the barrier rather than over it, and the apparent activation energy will be less than the barrier height. The solid lines of Fig. 8–1b indicate the region of phase space where mixing is significant and the broken lines where mixing is negligible. On the solid line mixing is weakest at the extreme ends of these lines and has its largest value at the top of the barrier. For strongly adiabatic situations, the solid line includes the potential minima corresponding to the reactants' and products' collision complexes. Then the nuclei have their normal geometry so that tunnelling might occur without significant alteration of nuclear geometry. Of course, the diagram greatly oversimplifies the situation. We would never expect all the nuclei to tunnel but some of the more energetically expensive nuclear distortions necessary to pass over the potential barrier may be bypassed in this way. If the nuclear tunnelling effect is large, calculations of the free energy of activation by the method of Marcus or one of the alternatives, all of which ignore tunnelling, may be seriously in error. To investigate the size of the tunnelling effect we can use the equations of Sutin and apply them to the reaction between aquoferric and aquoferrous ions.

Tunnelling can occur from each vibrational level of the reactants, each level thus making some reduction in the apparent activation energy no matter how small. The nuclear tunnelling potential barrier is shown in Fig. 8–1b as a composite of the potential energy curves for the breathing modes of the inner-shell water-ion vibrations for reactants I and the products II to conform with the discussion of Chapter 4. The activated-complex coordinate descriptions correspond to the top of the barrier and the solid potential barrier has a rather uncertain height determined by the interaction energy. Roughly, the line is solid when the interaction energy for the nuclear coordinates at each point on the line corresponds to a significant interaction energy. In contrast to the usual treatment of such problems, the tunnelling probability will depend not only on the total distance a given vibrational level lies below the top of the barrier but also on the amount of mixing of the zero-order surfaces at that level. The problem is not so complicated for homonuclear reactions but becomes considerably more so for heteronuclear reactions as a result of different degrees of mixing on the two sides of the barrier at a given height.

In Sutin's treatment, when all the contributions from the reactants' vibrational modes are summed, taking into account the Boltzmann distribution of populations in the vibrational states, a nuclear tunnelling factor Γ is introduced as a multiplicative factor in the rate constant. This factor is the ratio of the rate constant, calculated with tunnelling corrections, to that calculated without tunnelling, and is given approximately by

$$\Gamma = \frac{\exp\left(\Delta F_i^{\ddagger}/RT\right)}{RT} \int_0^\infty \exp\left(-\frac{W}{RT}\right) \exp\left(-\tfrac{4}{3}\gamma l\right) dW \qquad (8\text{--}1)$$

in which[189]

$$\exp\left(-\tfrac{4}{3}\gamma l\right) = \text{nuclear transmission coefficient.}$$
$$\gamma = (2\pi/h)\sqrt{2m(\Delta F_i^{\ddagger} - W)}.$$
$$W = \text{energy of the tunnelling mass, } m.$$
$$l = \text{width of the barrier at energy level } W.$$
$$= \frac{2(\Delta F_i^{\ddagger} - W)}{s}.$$
$$s = \text{slope of triangular potential-energy barrier.}$$

This equation does not adequately account for the poor mixing as the vibrational energies depart more and more from values at the top of the barrier. Thus it makes a considerable overestimate. The value of Γ given by Sutin[18] was approximately 40 for an iron-water distance of 2.09 Å at the summit of the potential barrier. However, the value of Γ is very sensitive to the tunnelling mass and to the slope, s, of the barrier, and these are uncertain quantities. For example, if we assume that tunnelling occurs from the $W = 0$ energy level only, that the tunnelling mass is six water molecules or 108 AMU, and that the slope is $\Delta F_i^{\ddagger}/0.04$ Å or 2.08×10^{-3} erg/cm. (since $\Delta F_i^{\ddagger} \simeq 12$ kcal./mole), then

$$\Gamma = e^{2.5}$$

However, if the slope is taken as $\Delta F_i^{\ddagger}/0.08$ Å, and all other quantities left unchanged, then

$$\Gamma = e^{-15.0}$$

indicating that tunnelling from the lowest energy level is a completely negligible process compared to activated complexes going over the barrier.

The nuclear tunnelling coefficient will be greater for the higher vibrational levels, but the number of reactant systems in the higher levels decreases exponentially according to the Boltzmann distribution. However, the nuclear tunnelling coefficient will increase more rapidly than the Boltzmann factor will decrease because of the higher power of W in the exponent of the former quantity so that the integrand in Eq. 8–1 will continually increase over the energy range $0 \leqslant W \leqslant \Delta F_i^{\ddagger}$. For $W \leqslant \Delta F_i^{\ddagger}$ the nuclear tunnelling coefficient will be essentially unity so that the integral in Eq. 8–1 will be no more important for this range of energies than it is when nuclear tunnelling is neglected.

The nuclear tunnelling factor contains the square root of m, the tunnelling mass, in the exponent. Because the water molecules in the outer coordination shells, as well as those in the inner coordination shells, of the ions must tunnel, the value of m is quite uncertain. An increase of a factor of two in this parameter will increase the exponent by approximately 40 percent and seriously affect the nuclear transmission coefficient.

Perhaps even more questionable than the values to be assigned to the slope of the barrier and to the tunnelling mass is the neglect of the decrease in the nuclear transmission coefficient as the nuclear configuration departs widely from that of the activated-complex region. As we have already noted, this neglect leads to the inclusion of the lower vibrational quantum numbers and to an overestimate of the tunnelling factor. In the treatment, we have ignored any explicit statement about the contributions to the activation energy from other ligand modes and solvent modes, but these are effectively included by the use of the experimental free energy of activation. Hence it is unlikely that nuclear tunnelling can introduce an error of detectable magnitude in the calculation of rate constants by the method of Marcus and Hush. There is one probable exception to this, and that is the special case of hydrogen. If the important modes undergoing distortion are such that hydrogen atom rearrangement occurs, considerably higher tunnelling factors may be expected.

References

1. W. F. LIBBY, "Theory of Electron Exchange Reactions in Aqueous Solution," *J. Phys. Chem.*, **56**, 863 (1952).

2. R. J. MARCUS, B. J. ZWOLINSKI, and H. EYRING, "The Electron Tunnelling Hypothesis for Electron Exchange Reactions," *J. Phys. Chem.*, **58**, 432 (1954).

3. B. J. ZWOLINSKI, R. J. MARCUS, and H. EYRING, "Inorganic Oxidation-Reduction Reactions in Solution," *Chem. Rev.*, **55**, 157 (1955).

4. J. WEISS, "On the Theory of Electron-Transfer Processes in Aqueous Solutions," *Proc. Roy. Soc., London*, **A222**, 128 (1954).

5. K. J. LAIDLER, "Some Theoretical Aspects of Electron-Transfer Processes in Aqueous Solution," *Can. J. Chem.*, **37**, 138 (1959).

6. R. A. MARCUS, "On the Theory of Oxidation-Reduction Reactions Involving Electron Transfer. I.," *J. Chem. Phys.*, **24**, 966 (1956).

7. R. A. MARCUS, "On the Theory of Oxidation-Reduction Reactions Involving Electron Transfer. II. Applications to Data on the Rates of Isotopic Exchange Reactions," *J. Chem. Phys.*, **26**, 867 (1957).

8. R. A. MARCUS, "On the Theory of Oxidation-Reduction Reactions Involving Electron Transfer. III. Applications to Data on the Rates of Organic Redox Reactions," *J. Chem. Phys.*, **26**, 872 (1957).

9. R. A. MARCUS, "Theory of Oxidation-Reduction Reactions Involving Electron Transfer. IV. A Statistical-Mechanical Basis for Treating Contributions from Solvent, Ligands and Inert salt," *Discussions Faraday Soc.*, No. 29, 21 (1960).

10. R. A. MARCUS, "On the Theory of Oxidation-Reduction Reactions Involving Electron Transfer. V. Comparison and Properties of Electro-Chemical and Chemical Rate Constants," *J. Phys. Chem.*, **67**, 853 (1963).

11. R. A. MARCUS, "On the Theory of Electron-Transfer Reactions. VI. Unified Treatment for Homogeneous and Electrode Reactions," *J. Chem. Phys.*, **43**, 679 (1965).

12. J. HALPERN and L. E. ORGEL, "The Theory of Electron Transfer between Metal Ions in Bridged Systems," *Discussions Faraday Soc.*, No. 29, 32 (1960).

13. V. G. LEVICH and R. R. DOGONADZE, "Adiabatic Theory for Electron-Transfer Processes in Solution," *Dokl. Akad. Nauk SSSR*, **113,**158 (1960) (English translation, *Proc. Acad. Sci. USSR, Phys. Chem. Sect.*, **133**, 591 [1960]); *Collection Czech. Chem. Commun.*, **26**, 193 (1961).

14. H. M. MCCONNELL, "Intramolecular Charge Transfer in Aromatic Free Radicals," *J. Chem. Phys.*, **35**, 508 (1961).

15. N. S. Hush, "Adiabatic Theory of Outer Sphere Electron-Transfer Reactions in Solution," *Trans. Faraday Soc.*, **57**, 557 (1961).

16. E. Sacher and K. J. Laidler, "A Theory of Non-Adiabatic Electron-Transfer in Aqueous Solution," *Trans. Faraday Soc.*, **59**, 396 (1963).

17. J. Halpern, "Mechanisms of Electron Transfer and Related Processes in Solution," *Quart. Rev., London*, **15**, 207 (1961).

18. N. Sutin, "Electron Exchange Reactions," *Ann. Rev. Nucl. Sci.*, **12**, 285 (1962).

19. R. A. Marcus, "Chemical and Electrochemical Electron-Transfer Theory," *Ann. Rev. Phys. Chem.*, **15**, 155 (1964).

20. H. Taube, H. Myers, and R. C. Rich, "Observations on the Mechanism of Electron Transfer in Solution," *J. Am. Chem. Soc.*, **75**, 4118 (1953).

21. H. Taube and H. Myers, "Evidence for a Bridged Activated Complex for Electron Transfer Reactions," *J. Am. Chem. Soc.*, **76**, 2103 (1954).

22. R. T. M. Fraser, D. K. Sebera, and H. Taube, "Substitution Coupled to Electron Transfer," *J. Am. Chem. Soc.*, **81**, 2906 (1959).

23. R. T. M. Fraser and H. Taube, "Activation of Bridging Groups in Electron Transfer. II. The Position of Bond-Breaking in Ester Hydrolysis," *J. Am. Chem. Soc.*, **81**, 5000 (1959).

24. R. T. M. Fraser and H. Taube, "Activation by Electron Transfer. Induced cis-trans Isomerism," *J. Am. Chem. Soc.*, **81**, 5514 (1959).

25. D. Bunn, F. S. Dainton, and S. Duckworth, "Redox Exchange Between Ferrous Ions and FeN_3^{+2} in Aqueous Solution," *Trans. Faraday Soc.*, **57**, 1131 (1961).

26. A. C. Wahl, "Rapid Electron-Transfer Isotope Exchange Reactions," *Z. Elektrochem.*, **64**, 90 (1960).

27. P. J. Zandstra and S. I. Weissman, "Effects of Ion Association on Rates of Oxidation-Reduction Transfer Reactions in the Naphthalene-Naphthalenide Systems," *J. Am. Chem. Soc.*, **84**, 4408 (1962).

28. W. L. Reynolds, "Rate of Electron Exchange Between 2,2′-Bipyridine and 2,2′-Bipyridine Negative Ion," *J. Phys. Chem.*, **67**, 2866 (1963).

29. R. A. Horne, "The Kinetics of the Oxalate Catalysis of the Iron(II)-Iron(III) Electron Exchange Reaction in Aqueous Solution," *J. Phys. Chem.*, **64**, 1512 (1960).

30. J. Halpern, R. J. Legare, and R. Lumry, "Measurement of the Rate of Electron Transfer Between Tris-(4,7-Dimethyl-1,10-Phenanthroline) Iron(II) and Hexachloroiridate(IV) by the T-Jump Method," *J. Am. Chem. Soc.*, **85**, 680 (1963).

31. R. L. Ward and S. I. Weissmann, "Electron Spin Resonance Study of the Electron Exchange between Naphthalene Negative Ion and Naphthalene," *J. Am. Chem. Soc.*, **79**, 2086 (1957).

32. R. W. Dodson and N. Davidson, *J. Phys. Chem.*, **56**, 866 (1952). Contribution to the discussion.

33. W. L. Reynolds and R. W. Lumry, "Role of Water in Oxidation-Reduction Reactions," *J. Chem. Phys.*, **23**, 2460 (1955).

34. R. A. Horne and E. H. Axelrod, "Proton Mobility and Electron Exchange in Aqueous Media," *J. Chem. Phys.*, **40**, 1518 (1964).

35. J. P. Hunt, *Metal Ions in Aqueous Solution*, W. A. Benjamin, Inc., New York, 1963.

36. R. A. Robinson and R. H. Stokes, *Electrolyte Solutions*, Butterworth Scientific Publications, London, 1959. Second ed.

37. R. W. Gurnee, *Ionic Processes in Solution*, McGraw-Hill Book Co. Inc., New York, 1953.

38. M. Kaminsky, "Ion-Solvent Interaction and the Viscosity of Strong-Electrolyte Solutions," *Discussions Faraday Soc.*, No. 24, 171 (1957).

39. O. Ya. Samoilov, "A New Approach to the Study of Hydration of Ions in Aqueous Solutions," *Discussions Faraday Soc.*, No. 24, 141 (1957).

40. H. S. Frank and Wen-Yang Wen, "Structural Aspects of Ion-Solvent Interaction in Aqueous Solution: A Suggested Picture of Water Structure," *Discussions Faraday Soc.*, No. 24, 133 (1957).

41. L. Benjamin and V. Gold, "A Table of Thermodynamic Functions of Ionic Hydration," *Trans. Faraday Soc.*, **50**, 797 (1954).

42. J. P. Hunt and H. Taube, "The Exchange of Water Between Hydrated Cations and Solvent" *J. Chem. Phys.*, **19**, 602 (1951).

43. F. Basolo and R. G. Pearson, *Mechanisms of Inorganic Reactions*, John Wiley & Sons, Inc., New York, 1958.

44. Circular of the National Bureau of Standards, No. 500, "Selected Values of Chemical Thermodynamic Properties."

45. Reference 43, Table 9, p. 66.

46. C. J. Ballhausen, *Introduction to Ligand Field Theory*, McGraw-Hill Book Co., Inc., New York, 1962.

47. L. E. Orgel, *An Introduction to Transition Metal Chemistry*, John Wiley & Sons, Inc., New York, 1960.

48. J. S. Griffith, "On the Stabilities of Transition Metal Complexes. II," *J. Inorg. Nucl. Chem.*, **2**, 229 (1956).

49. J. H. Van Vleck, "The Jahn-Teller Effect and Crystalline Stark Splitting for Clusters of the Form XY_6," *J. Chem. Phys.*, **7**, 72 (1939); D. Polder, "On the Theory of the Paramagnetic Anisotropy of Some Hydrated Cupric Salts" *Physica*, **9**, 709 (1942).

50. A. D. Liehr, Chapter in *Progress in Inorganic Chemistry*, Vol. 5, Interscience Publishers, Inc., New York, 1963. Edited by F. A. Cotton, p. 385.

51. K. J. Laidler, "The Entropies of Ions in Aqueous Solution. I. Dependence on Charge and Radius," *Can. J. Chem.*, **34**, 1107 (1956).

52. P. C. Scott and Z Z. Hugus, Jr., "Partial Molal Entropies of Ions in Aqueous Solution," *J. Chem. Phys.*, **27**, 1421 (1957).

53. K. J. Laidler, "Comments on 'Partial Molal Entropies of Ions in Aqueous Solution' by P. C. Scott and Z Z. Hugus," *J. Chem. Phys.*, **27**, 1423 (1957).

54. L. Pauling, *The Nature of the Chemical Bond*, Cornell University Press, Ithaca, New York, 1960. Third edition, p. 505–511.

55. H. A. C. McKay, "Kinetics of Some Exchange Reactions of the Type $RI + {}^{*}I^{-} \rightleftarrows R{}^{*}I + I^{-}$ in Alcoholic Solution," *J. Am. Chem. Soc.*, **65**, 702 (1943); Nature, **42**, 997 (1938).

56. R. J. PRESTWOOD and A. C. WAHL, "The Kinetics of the Thallium(I)-Thallium(III) Exchange Reaction," *J. Am. Chem. Soc.*, **71**, 3137 (1949).

57. L. VAN ALTEN and C. N. RICE, "Exchange Reaction Between Ferric and Ferrous Ions in Perchloric Acid Using a Diffusion Separation Method," *J. Am. Chem. Soc.*, **70**, 883 (1948).

58. V. J. LINNENBOM and A. C. WAHL, "Exchange Reactions Between Cerium(III) and Cerium(IV) and Between Iron(II) and Iron(III)," *J. Am. Chem. Soc.*, **71**, 2589 (1949).

59. H. A. KIERSTEAD, "Ferrous-Ferric Electron Transfer Reaction in Perchloric Acid Solution," *J. Chem. Phys.*, **18**, 756 (1950).

60. J. SILVERMAN and R. W. DODSON, "The Exchange Reaction Between the Two Oxidation States of Iron in Acid Solution," *J. Phys. Chem.*, **56**, 846 (1952).

61. S. FUKUSHIMA and W. L. REYNOLDS, "D_2O Effect on $\Delta H^{\ddagger}$ and $\Delta S^{\ddagger}$ in the Iron(II)-Iron(III) Electron-Exchange Reaction," *Talanta*, **11**, 283 (1964).

62. J. F. BELOW, Jr., R. E. CONNICK, and C. P. COPPEL, "Kinetics of the Formation of the Ferric Thiocyanate Complex," *J. Am. Chem. Soc.*, **80**, 2961 (1958).

63. R. E. CONNICK and C. P. COPPEL, "Kinetics of the Formation of the Ferric Chloride Complex," *J. Am. Chem. Soc.*, **81**, 6389 (1959).

64. D. POULI and W. MacF. SMITH, "The Kinetics of the Formation of the Monofluoro Complex of Iron(III) in Aqueous Solution," *Can. J. Chem.*, **38**, 567 (1960).

65. G. S. LAURENCE, "The Kinetics of the Exchange Reactions between Ferrous Ions and the Ferric Thiocyanate Complexes," *Trans. Faraday Soc.*, **53**, 1326 (1957).

66. J. HUDIS and A. C. WAHL, "The Kinetics of the Exchange Reactions between Iron(II) Ion and the Fluoride Complexes of Iron(III)," *J. Am. Chem. Soc.*, **75**, 4153 (1953).

67. J. MENASHI, S. FUKUSHIMA, C. FOXX, and W. L. REYNOLDS, "The Role of FeF^{+2} in the $Fe^{+2} + Fe^{+3} + F^-$ Isotope Exchange Reaction," *Inorg. Chem.*, **3**, 1242 (1964).

68. G. DULZ and N. SUTIN, "The Effect of Chloride Ions on the Kinetics of the Oxidation of Chromium(II) by Iron(III)," *J. Am. Chem. Soc.*, **86**, 829 (1964).

69. R. J. CAMPION, T. J. CONOCCHIOLI, and N. SUTIN, "The Inner-Sphere Activated Complex for the Electron Exchange of Iron(II) and the Monochloro Complex of Iron(III)," *J. Am. Chem. Soc.*, **86**, 4591 (1964).

70. R. A. HORNE, Ph.D. thesis, Columbia University, New York, 1955.

71. J. C. SHEPPARD and L. C. BROWN, "The Effect of Several Oxy-Acids on the Rate of Electron Transfer Between Iron(II) and Iron(III) Ions in Perchloric Acid," *J. Phys. Chem.*, **67**, 1025 (1963).

72. W. L. REYNOLDS and S. FUKUSHIMA, "Iron(II) + Iron(III) Isotope Exchange in Presence of Sulfate Ions," *Inorg. Chem.*, **2**, 176 (1963).

73. R. L. S. WILLIX, "Ferrous-Ferric Redox Reaction in the Presence of Sulfate Ion," *Trans. Faraday Soc.*, **59**, 1315 (1963).

74. K. Bächmann and K. H. Lieser, "Der Elektronenaustausch zwischen Fe(II) und Fe(III) in Anwesenheit von Sulfationen," *Z. Physik. Chem., N. F.,* **36**, 236 (1963).

75. A. McAuley and C. H. Brubaker, Jr., "The Acceleration of the Iron(II)-(III) Exchange by Tartaric Acid," *Inorg. Chem.,* **3**, 273 (1964).

76. M. Dietrich, Ph.D. thesis, Washington University, St. Louis, Mo., 1961.

77. B. N. Mattoo, "Stability of Metal Complexes in Solution. III. Ion Association in Ferric Sulfate and Nitrate Solutions at Low Fe(III) Concentrations," *Z. Physik. Chem., N. F.,* **19**, 156 (1959).

78. J. Hudis and R. W. Dodson, "Rate of Ferrous-Ferric Exchange in D_2O," *J. Am. Chem. Soc.,* **78**, 911 (1956).

79. N. Sutin, J. K. Rowley, and R. W. Dodson, "Chloride Complexes of Iron(III) Ions and the Kinetics of the Chloride Catalyzed Exchange Reaction between Iron(II) and Iron(III) in Light and Heavy Water," *J. Phys. Chem.,* **65**, 1248 (1961).

80. A. Zwickel and H. Taube, "The Deuterium Isotope Effect for an Oxidation-Reduction Reaction between Aquo Ions," *J. Am. Chem. Soc.,* **81**, 1288 (1959).

81. A. E. Ogard and H. Taube, "Halides as Bridging Groups for Electron Transfer in the Systems $Cr^{+2} + (NH_3)_5CrX^{+2}$," *J. Am. Chem. Soc.,* **80**, 1084 (1958).

82. R. A. Horne, "Kinetics of the Iron(II)-Iron(III) Electron Exchange Reaction in Ice Media," *J. Inorg. Nucl. Chem.,* **25**, 1139 (1963).

83. N. Sutin, "The Absorption Spectrum of Ferric Perchlorate and the Rate of the Ferrous-Ferric Exchange Reaction in Isopropyl Alcohol," *J. Phys. Chem.,* **64**, 1766 (1960).

84. A. G. Maddock, "Ferrous-Ferric Exchange in Non-Aqueous Solvents," *Trans. Faraday Soc.,* **55**, 1268 (1959).

85. J. Menashi, W. L. Reynolds, and G. Van Auken, "Iron(II) + Iron(III) Isotope Exchange in Dimethyl Sulfoxide," *Inorg. Chem.,* **4**, 299 (1965).

86. W. L. Reynolds, N. Liu, and J. Mickus, "The Exchange of Iron between the Aquo Ferrous and Ferric Versenate Ions," *J. Am. Chem. Soc.,* **83**, 1078 (1961).

87. M. H. Ford-Smith and N. Sutin, "The Kinetics of the Reactions of Substituted 1,10-Phenanthroline, 2,2'-Dipyridine and 2,2'-,2"-Tripyridine Complexes of Iron(III) with Iron(II) Ions," *J. Am. Chem. Soc.,* **83**, 1830 (1961).

88. N. Sutin and B. M. Gordon, "The Kinetics of the Oxidation of the Iron(II) Ion by the Tris-(1,10-Phenanthroline)-Iron(III) Ion," *J. Am. Chem. Soc.,* **83**, 70 (1961).

89. E. Eichler and A. C. Wahl, "Electron-Exchange Reactions Between Large Complex Cations," *J. Am. Chem. Soc.,* **80**, 4145 (1958).

90. B. M. Gordon, L. L. Williams, and N. Sutin, "The Kinetics of the Oxidation of Iron(II) Ions and of Coordination Complexes," *J. Am. Chem. Soc.,* **83**, 2061 (1961).

91. A. C. WAHL and C. F. DECK, "Rate of the Ferrocyanide-Ferricyanide Exchange Reaction," *J. Am. Chem. Soc.*, **76**, 4054 (1954); A. C. WAHL, "Rapid Electron-Transfer Isotope-Exchange Reactions," *Z. Electrochem.*, **64**, 90 (1960).

92. H. TAUBE, "Mechanisms of Redox Reactions of Simple Chemistry," *Advan. Inorg. Chem. Radiochem.*, **1**, 1 (1959).

93. A. ZWICKEL and H. TAUBE, "Kinetics of Some Electron Transfer Reactions of Cobalt(III)," *J. Am. Chem. Soc.*, **83**, 793 (1961).

94. J. F. ENDICOTT and H. TAUBE, "Kinetics of Some Outer-Sphere Electron-Transfer Reactions," *J. Am. Chem. Soc.*, **86**, 1686 (1964).

95. J. P. CANDLIN, J. HALPERN, and D. C. TRIMM, "Kinetics of the Reduction of Some Cobalt(III) Complexes by Chromium(II), Vanadium(II), and Europium(II)," *J. Am. Chem. Soc.*, **86**, 1019 (1964).

96. H. S. GATES and E. L. KING, "A Study of the Equilibria in Acidic Chromium(III) Chloride Solutions," *J. Am. Chem. Soc.*, **80**, 5011 (1958).

97. A. W. ADAMSON and K. S. VORRES, "Charge Transfer Rates Between the Ethylenediaminetetraacetate Complexes of Co(II) and Co(III) and of Fe(II) and Fe(III)," *J. Inorg. Nucl. Chem.*, **3**, 206 (1956).

98. S. S. JONES and F. A. LONG, "Complex Ions from Iron and Ethylenediaminetetraacetate: General Properties and Radioactive Exchange," *J. Phys. Chem.*, **56**, 25 (1952).

99. L. EIMER and A. I. MEDALIA, "Exchange between the Tris-(5,6-dimethyl-1,10-phenanthroline) Complexes of Iron(II) and -(III)," *J. Am. Chem. Soc.*, **74**, 1592 (1952)).

100. T. S. LEE, I. M. KOLTHOFF, and D. L. LEUSSING, "Reaction of Ferrous and Ferric Ions with 1,10-Phenanthroline. II. Kinetics of Formation and Dissociation of Ferrous Phenanthroline," *J. Am. Chem. Soc.*, **70**, 3596 (1948).

101. N. MATAGA and W. L. REYNOLDS, unpublished work.

102. G. DULZ and N. SUTIN, "The Kinetics of the Oxidation of Iron(II) and Its Substituted Tris-(1,10-Phenanthroline) Complexes by Cerium(IV)," *Inorg. Chem.*, **2**, 917 (1963).

103. R. C. THOMPSON, "Some Exchange Experiments Involving Hexacyanoferrate(II) and Hexacyanoferrate(III) Ions," *J. Am. Chem. Soc.*, **70**, 1045 (1948).

104. Ch. HAENNY and G. ROCHAT, "Échanges isotopiques du fer. I.", *Helv. Chim. Acta*, **32**, 2441 (1949).

105. J. W. COBBLE and A. W. ADAMSON, "The Electron Transfer Exchange Reaction of Ferricyanide and Ferrocyanide Ions," *J. Am. Chem. Soc.*, **72**, 2276 (1950).

106. P. KING, C. F. DECK, and A. C. WAHL, paper presented at American Chemical Society meeting, St Louis, Mo., March 21–30, 1961.

107. S. I. WEISSMAN and M. COHN, "Spin Density in Octocyanomolybdate(V)," *J. Chem. Phys.*, **27**, 1440 (1957).

108. J. C. HINDMAN, J. C. SULLIVAN, and D. COHEN, "Kinetics of Reactions between Neptunium Ions. The Neptunium(IV)-Neptunium(VI) Reaction in Perchlorate Solution," *J. Am. Chem. Soc.*, **76**, 3278 (1954).

109. J. C. HINDMAN, J. C. SULLIVAN, and D. COHEN, "The Effect of Deuterium on the Kinetics of Reactions Involving Neptunium(IV), (V) and (VI) Ions," *J. Am. Chem. Soc.*, **81**, 2316 (1959).

110. J. C. SULLIVAN, D. COHEN, and J. C. HINDMAN, "Kinetics of Reactions Involving Neptunium(IV), Neptunium(V) and Neptunium(VI) Ions in Sulfate Media," *J. Am. Chem. Soc.*, **79**, 4029 (1957).

111. J. C. SULLIVAN and J. C. HINDMAN, "The Hydrolysis of Neptunium(IV)," *J. Am. Chem. Soc.*, **63**, 1332 (1959).

112. D. COHEN and J. C. HINDMAN, "The Neptunium(IV)-Neptunium(V) Couple in Perchloric Acid. The Partial Molal Heats and Free Energies of Formation of Neptunium Ions," *J. Am. Chem. Soc.*, **74**, 4682 (1952).

113. J. C. SULLIVAN, D. COHEN, and J. C. HINDMAN, "Isotopic Exchange Reactions of Neptunium Ions in Solution. II. The Np(IV)-Np(V) Exchange," *J. Am. Chem. Soc.*, **76**, 4275 (1954).

114. B. B. CUNNINGHAM and J. C. HINDMAN, in *The Actinide Elements*, McGraw-Hill Book Co., Inc., New York, 1954. Edited by G. T. Seaborg and J. J. Katz.

115. L. H. JONES and R. A. PENNEMAN, "Infrared Spectra and Structure of Uranyl and Transuranium(V) and (VI) Ions in Aqueous Perchloric Acid Solution," *J. Chem. Phys.*, **21**, 542 (1953).

116. H. TAUBE "Anions as Bridging and Non-Bridging Ligands in Reactions of Co(III) Compounds with Cr^{+2}," *J. Am. Chem. Soc.*, **77**, 4481 (1955).

117. R. L. CARLIN, and J. O. EDWARDS, "Reaction of Thiocyanatopentaam-minecobalt(III) Ion with Chromous Ion," *J. Inorg. Nucl. Chem.*, **6**, 217 (1958).

118. R. K. MURMANN, H. TAUBE, and F. A. POSEY, "Mechanisms of Electron Transfer in Aquo Cations—The Reaction of RH_2O^{+3} with Cr^{+2}," *J. Am. Chem. Soc.*, **79**, 262 (1957).

119. D. K. SEBERA and H. TAUBE, "Organic Anions as Bridging Groups in Oxidation-Reduction Reactions," *J. Am. Chem. Soc.*, **83**, 1785 (1961).

120. R. T. M. FRASER, "Adjacent and Remote Attack in Electron Transfer Through Conjugated Ligands," *J. Am. Chem. Soc.*, **83**, 564 (1961).

121. G. SVATOS and H. TAUBE, "Malonate as a Bridging Group for Electron Transfer," *J. Am. Chem. Soc.*, **83**, 4172 (1961).

122. J. E. EARLEY and J. H. GORBITZ, "Mechanism of an Oxidation of Cr(II) in the Presence of Pyrophosphate," *J. Inorg. Nucl. Chem.*, **25**, 306 (1963).

123. W. KRUSE and H. TAUBE, "The Transfer of Oxygen in Some Oxidation-Reduction Reactions in Aquo Complexes," *J. Am. Chem. Soc.*, **82**, 526 (1960).

124. A. ANDERSON and N. A. BONNER, "The Exchange Reaction between Chromous and Chromic Ions in Perchloric Acid Solution," *J. Am. Chem. Soc.*, **76**, 3826 (1954).

125. R. T. M. FRASER, "Recent Advances in Electron-Transfer Reactions," *Rev. Pure Appl. Chem.*, **11**, 64 (1961).

126. R. T. M. FRASER, "Alkyl-Oxygen Fission of Ester Ligands in Electron-Transfer Reactions," *Proc. Chem. Soc.*, 317 (1960).

127. R. T. M. FRASER, "Conjugation and Mobile Bond Order in Electron Transfer Reactions," *J. Am. Chem. Soc.*, **83**, 4920 (1961); "Factors Controlling the Rate of Electron Transfer. I. The Effect of Halogen Substitution in Organic Mediators," *J. Am. Chem. Soc.*, **84**, 3436 (1962).

128. H. TAUBE, Symposium on "Mechanisms of Inorganic Reactions," June 22–24, 1964, at Lawrence, Kansas.

129. K. D. KOPPLE and G. F. SVATOS, "The Isomerization of Trialkylacetic Acids in Sulfuric Acid," *J. Am. Chem. Soc.*, **82**, 3227 (1960).

130. R. T. M. FRASER and H. TAUBE, "Remote Attack and Ester Hydrolysis on Electron Transfer," *J. Am. Chem. Soc.*, **83**, 2239 (1961).

131. R. T. M. FRASER and H. TAUBE, "Activation Effects and Rates of Electron Transfer," *J. Am. Chem. Soc.*, **83**, 2242 (1961).

132. R. T. M. FRASER, "The Effect of Chelation by Nonbridging Ligands on the Rate of Reduction of Co(III) Sulfato and Acetato Complexes," *Inorg. Chem.*, **2**, 954 (1963).

133. A. M. ZWICKEL and H. TAUBE, "The Rates and Mechanism of Reactions of $Cr(bip)_3^{+2}$ with Co(III) Complexes," *Discussions Faraday Soc.*, No. 29, 42 (1960).

134. T. M. DUNN, in *Modern Coordination Chemistry*, Interscience Publishers, Inc., New York, 1960, p. 256. Edited by J. Lewis and R. G. Wilkins.

135. J. P. CANDLIN, J. HALPERN, and S. NAKAMURA, "Inner- and Outer-Sphere Mechanisms in the Oxidation of Pentacyanocobaltate(II) by Penta-amminecobalt(III) Complexes," *J. Am. Chem. Soc.*, **85**, 2517 (1963).

136. W. M. LATIMER, *The Oxidation States of Elements and Their Potentials in Aqueous Solutions*, Prentice-Hall, Inc., Englewood Cliffs, N.J., 1952.

137. H. TAUBE and E. L. KING, "The Bridged Activated Complex for the Electron Exchange of Chromium(II) and Monochlorochromium(III) Ion," *J. Am. Chem. Soc.*, **76**, 4053 (1954).

138. D. L. BALL and E. L. KING, "The Exchange Reactions of Chromium(II) Ion and Certain Chromium(III) Complex Ions," *J. Am. Chem. Soc.*, **80**, 1091 (1958).

139. R. SNELLGROVE and E. L. KING, "Kinetics of Exchange of Chromium(II) Ion and Azidopentaaquochromium(III) Ion," *Inorg. Chem.*, **3**, 288 (1964).

140. YUAN-TSAN CHIA and E. L. KING, "The Reactions of Chromium(II) and the Isomeric Difluorochromium(III) Ions," *Discussions Faraday Soc.*, No. 29, 109 (1960).

141. R. SNELLGROVE and E. L. KING, "The Exchange of Chromium(II) Ion and *cis*-diazidotetraaquochromium(III) Ion. A Double-Bridged Transition State," *J. Am. Chem. Soc.*, **84**, 4609 (1962).

142. J. P. HUNT and J. E. EARLEY, "The Effect of Some Non-Bridging Ligands on the Cr(II)-Cr(III) Oxidation," *J. Am. Chem. Soc.*, **82**, 5312 (1960).

143. S. GLASSTONE, K. J. LAIDLER, and H. EYRING, *The Theory of Rate Processes*, McGraw-Hill Book Co., Inc., New York, 1941.

144. R. C. TOLMAN, *The Principles of Statistical Mechanics*, Oxford University Press, New York, 1938, pp. 412–4.

145. S. A. RICE, Personal communication to R. Lumry.

146. L. Landau, "Zur Theorie der Energieübertragung. II.," *Physik. Z., Sowjetunion*, **2**, 46 (1932).

147. C. Zener, "Non-Adiabatic Crossing of Energy Levels," *Proc. Roy. Soc. London*, **A137**, 696 (1932); "Dissociation of Excited Diatomic Molecules by External Perturbations,"*Proc. Roy. Soc. London*, **A140**, 660 (1933).

148. T. Wu and T. Omura, *Quantum Theory of Scattering*, Prentice-Hall, Inc., Englewood Cliffs, N.J., 1962.

149. H. S. W. Massey, "Theory of Atomic Collisions," *Handbuch der Physik*, **36**, 232 (1956).

150. D. M. Hercules, "Chemiluminescence Resulting from Electro-chemically Generated Species," *Science*, **145**, 808 (1964).

151. E. A. Chandross and F. I. Sonntag, "A Novel Chemiluminescent Electron Transfer Reaction," *J. Am. Chem. Soc.*, **86**, 3179 (1964).

152. E. F. Gurnee and J. L. Magee, "Interchange of Charge Between Gaseous Molecules in Resonant and Near-Resonant Processes," *J. Chem. Phys.*, **26**, 1237 (1957).

153. K. J. Laidler, *Chemical Kinetics*, McGraw-Hill Book Co. Inc., New York, 1950, pp. 385–86.

154. P. George and C. J. Stratmann, "The Oxidation of Myoglobin to Metmyoglobin by Oxygen," *Biochem. J.*, **51**, 418 (1954).

155. E. Antonini, "Interrelationship Between Structure and Function in Hemoglobin and Myoglobin," *Physiol. Rev.* **45**, 123 (1965).

156. J. Brooks, "The Oxidation of Haemoglobin to Methaemoglobin by Oxygen. II. The Relation between the Rate of Oxidation and the Partial Pressure of Oxygen," *Proc. Roy. Soc. London*, **B118**, 560 (1935).

157. Q. H. Gibson, "The Kinetics of Reaction between Haemoglobin and Gases," *Prog. Biophys. Biophys. Chem.*, **9**, 1 (1959).

158. N. Sutin, "Rate of Oxidation of Ferrohaemoglobin by Ferricyanide Ions," *Nature*, **190**, 438 (1961).

159. L. D. Landau and E. M. Lifshitz, *Quantum Mechanics*, Addison-Wesley Publishing Co., Inc., Reading, Mass. (1958), pp. 304–12.

160. N. F. Mott and H. S. W. Massey, *The Theory of Atomic Collisions*, Oxford University Press, London, 2nd Ed. (1949). Chapters 8 and 12.

161. C. A. Coulson and K. Zalewski, "Internal Conversion and the Crossing of Molecular Potential Energy Surfaces," *Proc. Roy. Soc. London*, **A268**, 437 (1962).

162. D. R. Bates, "Collisions Involving the Crossing of Potential Energy Curves," *Proc. Roy. Soc. London*, **A257**, 22 (1960).

163. E. E. Nikitin, "Non-Adiabatic Transitions Near the Turning Point in Atomic Collisions," *Opt. Spectros. (USSR) English Transl.*, **11**, 246 (1961).

164. W. Kauzmann, *Quantum Chemistry*, Academic Press, Inc., New York, 1957, pp. 539–41.

165. V. G. Levich and R. R. Dogonadze, "Theory of Radiationless Electron Transitions Between Ions in Solutions," *Dokl. Akad. Nauk SSSR*, **124**, 123 (1959) (*Proc. Acad. Sci. USSR, Phys. Chem. Sect., English Transl.*, **124**, 9 [1959]).

166. S. I. PEKAR, *Untersuchungen über die Elektronentheorie der Kristalle*, Akademie-Verlag, Berlin (1954).

167. R. R. DOGONADZE, "The Rate of Electron Exchange Reactions in Solutions," *Dokl. Akad. Nauk. SSSR*, **133**, 1368 (1960) (*Proc. Acad. Sci. USSR, Phys. Chem. Sect., English Transl.*, **133**, 765 [1960]).

168. R. R. DOGONADZE, "Semi-Classical Treatment of Electron Exchange Reactions in Solution," *Dokl. Akad. Nauk. SSSR*, **142**, 1108 (1961) (*Proc. Acad. Sci. USSR, Phys. Chem. Sect., English Transl.*, **142**, 156 [1961]).

169. C. A. COULSON and H. C. LONGUET-HIGGINS, "The Electronic Structure of Conjugated Systems. I. General Theory," *Proc. Roy. Soc. London*, **A191**, 39 (1947).

170. F. C. COLLINS and G. E. KIMBALL, "Diffusion-Controlled Reaction Rates," *J. Colloid Sci.*, **4**, 425 (1949).

171. A. WELLER, "Generalized Theory of Diffusion-Controlled Reactions and its Use in Fluorescent Solutions," *Z. Phys. Chem., N. F.*, **13**, 335 (1957).

172. R. M. NOYES, "Effects of Diffusion Rate on Chemical Kinetics," *Prog. Reaction Kinetics*, **1**, 129 (1961).

173. P. DEBYE, "Reaction Rates in Ionic Solutions," *Trans. Electrochem. Soc.*, **82**, 265 (1942).

174. R. LAFONT, "Vibrations des ions hexahydrates $M(H_2O)_6$ dans les cristaux des sulfates orthorhombiques à sept molecules d'eau," *Compt. rend.*, **244**, 1481 (1957).

175. R. W. LUMRY and H. EYRING, "Conformation Changes of Proteins," *J. Phys. Chem.*, **58**, 110 (1954).

176. H. DIEBLER and N. SUTIN, "The Kinetics of Some Oxidation-Reduction Reactions Involving Manganese(III)," *J. Phys. Chem.*, **68**, 174 (1964).

177. R. J. CAMPION, N. PURDIE, and N. SUTIN, "The Kinetics of Some Related Electron-Transfer Reactions," *Inorg. Chem.*, **3**, 1091 (1964).

178. R. A. MARCUS, "Generalization of the Activated Complex Theory of Reaction Rates. I. Quantum Mechanical Treatment," *J. Chem. Phys.*, **41**, 2614 (1964); "II. Classical Mechanical Treatment," *ibid.*, **41**, 2624 (1964).

179. N. TANAKA and R. TAMAMUSHI, "Kinetic Parameters of Electrode Reactions" (a report presented to the Commission on Electrochemical Data of the Section of Analytical Chemistry of I.U.P.A.C., at the International Congress of Pure and Applied Chemistry, Montreal, 1961). Copies are obtainable from H. Fischer, Department of Electrochemistry, Institute of Technology, Karlsruhe, Germany.

180. A. F. JOFFE, "Two Mechanisms for the Motion of Free Charges," *Fiz. Tver. Tela.* **1**, 157, 160 (1959) (*Soviet Phys. Solid State*, **1**, 139, 141 [1959]).

181. F. J. MORIN, "Oxides of the 3d Transition Metals," in *Semiconductors*, Reinhold Publishing Corp., New York, 1959, ed. N. B. Hannay, pp. 600–33.

182. R. A. MARCUS, "A Theory of Electron Transfers at Electrodes," paper presented at 14th Meeting, International Commission Electrochemical Thermodynamics and Kinetics, Moscow (1963).

183. N. S. Hush, "Adiabatic Rate Processes at Electrodes, I. Energy-Charge Relationships," *J. Chem. Phys.*, **28**, 962 (1958); cf. *Z. Elektrochem.*, **61**, 734 (1957).

184. J. F. Dewald, "Semi-Conductor Electrodes," in *Semiconductors*, Reinhold Publishing Co., New York, 1959, ed. N. B. Hannay, pp. 727–52.

185. H. Gerischer, "On the Mechanism of Redox-Reactions on Metals and on Semi-Conductors. I. General Considerations of Electron-Transfer Between a Solid and a Redox Electrolyte," *Z. Physik. Chem. Frankfurt*, **26**, 223 (1960); "II. Metal Electrodes," *Z. Physik. Chem. Frankfurt*, **26**, 325 (1960); "III. Semi-Conductor Electrodes," *Z. Physik. Chem. Frankfurt*, **27**, 48 (1961).

186. J. E. B. Randles, "Kinetics of Rapid Electrode Reactions. II. Rate Constants and Activation Energies of Electrode Reactions," *Trans. Faraday Soc.*, **48**, 828 (1952).

187. R. R. Dogonadze and Y. A. Chizmadzhev, "Calculations of the Possibility of the Elementary Act of Some Heterogeneous Oxidation-Reduction Reactions," *Dokl. Akad. SSSR*, **144**, 1077 (1962) (*Proc. Acad. Sci. USSR, Phys. Chem. Sect., English Transl.*, **144**, 463 [1962]); "The Kinetics of Certain Electrochemical Oxidation-Reduction Reactions of Metals," *ibid.*, **145**, 848 (1962) (*ibid.*, **145**, 563 [1962]).

188. A. A. Vlcek, *Sixth International Conference on Coordination Chemistry*, The Macmillan Co., New York, 1961, ed. S. Kirschner, pp. 590–603.

189. N. F. Mott and I. N. Sneddon, *Wave Mechanics and Its Applications*, Clarendon Press, Oxford, 1948, Chap. I.

Index